T0305872

Industrial Internet of Things Security

The industrial landscape is changing rapidly, and so is global society. This change is driven by the growing adoption of the Industrial Internet of Things (IIoT) and artificial intelligence (AI) technologies. IIoT and AI are transforming the way industrial engineering is done, enabling new levels of automation, productivity, and efficiency. However, as IIoT and AI become more pervasive in the industrial world, they also offer new security risks that must be addressed to ensure the reliability and safety of critical systems.

Industrial Internet of Things Security: Protecting AI-Enabled Engineering Systems in Cloud and Edge Environments provides a comprehensive guide to IIoT security, covering topics such as network architecture, risk management, data security, and compliance. It addresses the unique security challenges that the cloud and edge environments pose, providing practical guidance for securing IIoT networks in these contexts. It includes numerous real-world case studies and examples, providing readers with practical insights into how IIoT security and AI-enabled industrial engineering are being implemented in various industries. Best practices are emphasized for the readers to ensure the reliability, safety, and security of their systems while also learning the latest developments in IIoT security for AI-enabled industrial engineering systems in this rapidly evolving field.

By offering step-by-step guidance for the implantation process along with best practices, this book becomes a valuable resource for practitioners and engineers in the areas of industrial engineering, IT, computer engineering, and anyone looking to secure their IIoT network against cyber threats.

Intelligent Manufacturing and Industrial Engineering

Series Editor: Ahmed A. Elngar, Beni-Suef Uni.
Mohamed Elhoseny, Mansoura University, Egypt

Machine Learning Adoption in Blockchain-Based Intelligent Manufacturing
Edited by Om Prakash Jena, Sabyasachi Pramanik, Ahmed A. Elngar

Integration of AI-Based Manufacturing and Industrial Engineering Systems with the Internet of Things
Edited by Pankaj Bhambri, Sita Rani, Valentina E. Balas, and Ahmed A. Elngar

AI-Driven Digital Twin and Industry 4.0: A Conceptual Framework with Applications
Edited by Sita Rani, Pankaj Bhambri, Sachin Kumar, Piyush Kumar Pareek, and Ahmed A. Elngar

Technology Innovation Pillars for Industry 4.0: Challenges, Improvements, and Case Studies
Edited by Ahmed A. Elngar, N. Thillaiarasu, T. Saravanan, and Valentina Emilia Balas

Internet of Things and Big Data Analytics-Based Manufacturing
Edited by Arun Kumar Rana, Sudeshna Chakraborty, Pallavi Goel, Sumit Kumar Rana, and Ahmed A. Elngar

Industrial Internet of Things Security: Protecting AI-Enabled Engineering Systems in Cloud and Edge Environments
Edited by Sunil Kumar Chawla, Neha Sharma, Ahmed A. Elngar, Prasenjit Chatterjee, and P. Naga Srinivasu

For more information about this series, please visit: https://www.routledge.com/Mathematical-Engineering-Manufacturing-and-Management-Sciences/book-series/CRCIMIE

Industrial Internet of Things Security

Protecting AI-Enabled Engineering Systems in Cloud and Edge Environments

Edited by Sunil Kumar Chawla, Neha Sharma,
Ahmed A. Elngar, Prasenjit Chatterjee and
P. Naga Srinivasu

CRC Press
Taylor & Francis Group
Boca Raton London New York

CRC Press is an imprint of the
Taylor & Francis Group, an **informa** business

Designed cover image: Shutterstock - TenPixels

First edition published 2025
by CRC Press
2385 NW Executive Center Drive, Suite 320, Boca Raton, FL 33431

and by CRC Press
4 Park Square, Milton Park, Abingdon, Oxon, OX14 4RN

CRC Press is an imprint of Taylor & Francis Group, LLC

ISBN: 978-1-032-73850-5 (hbk)
ISBN: 978-1-032-73854-3 (pbk)
ISBN: 978-1-003-46628-4 (ebk)

DOI: 10.1201/9781003466284

Typeset in Times
by KnowledgeWorks Global Ltd.

Contents

Preface

The industrial landscape is changing rapidly, driven by the growing adoption of the Industrial Internet of Things (IIoT) and artificial intelligence (AI) technologies. IIoT and AI are transforming the way industrial engineering is done, enabling new levels of automation, productivity, and efficiency. However, as IIoT and AI become more pervasive in the industrial world, they also introduce new security risks that must be addressed to ensure the reliability and safety of critical systems.

This book is designed to provide a comprehensive and practical guide to securing IIoT networks and implementing AI-enabled industrial engineering systems in cloud and edge environments. We believe that this book will be a valuable resource for engineers, practitioners, and researchers who are interested in IIoT security and AI-enabled industrial engineering.

In this book, we start with an overview of IIoT and AI in industrial engineering, and discuss the unique security challenges posed by cloud and edge environments. We then delve into the key topics of IIoT security, including network architecture, risk management, data security, and compliance. We also explore the benefits and challenges of AI in industrial engineering and provide best practices for securing AI-enabled industrial engineering systems

Throughout the book, we provide real-world examples to help readers understand how these technologies can be applied in various industries and contexts. We also provide practical guidance and checklists to help readers implement IIoT security and AI-enabled industrial engineering systems effectively

We hope that this book will help readers gain a deeper understanding of IIoT security and AI-enabled industrial engineering, and provide them with the knowledge and tools they need to secure their IIoT networks and optimize their industrial processes.

Chapter 1 examines unique security requirements of cloud and edge computing, as well as the specific vulnerabilities introduced by AI systems.

Chapter 2 presents security architecture for industrial systems having automation through the integration of Internet of Things (IoT) and smart sensors. The authors have studied layer-wise requirements of IIoT-based systems followed by proposing various security protection strategies.

Chapter 3 explores risk management in industrial systems. The chapter highlights the necessity for a comprehensive risk management framework meticulously tailored to the intricacies of IIoT. It also emphasizes the pivotal role of cybersecurity measures, including network segmentation, access control, and threat detection, in the prevention of malicious activities. Additionally, the chapter delves into the implications of observing regulation compliance, considering data privacy concerns, and complying with industry standards within the realm of risk management in industrial systems.

Chapter 4 offers an essential perspective of dealing with data security for industrial systems. The chapter provides a comprehensive overview of various approaches to IIoT security that involve robust encryption and authentication, network segmentation and isolation, device management, anomaly detection, intrusion prevention systems, regulatory compliance, employee training, and incident response planning.

Chapter 5 delves into the realm of compliance within the IIoT landscape, dissecting various facets crucial for navigating regulatory frameworks and ensuring adherence to standards.

Chapter 6 proposes a lightweight secure key authentication scheme for IoT. It gives a fully structured group key negotiation protocol using the underlying mathematical structure of balanced incomplete block designs and elliptic curve Qu-Vanstone (ECQV) authentication protocol. The protocol ensures resistance against chosen plaintext attacks while allowing flexible adaptation of the number of participants in group key negotiation protocols, thereby enhancing both the security and efficiency of the protocol.

Chapter 7 drives a deep learning approach for enhancing security in industrial IoT-based systems. This study employs deep learning techniques to increase secret capacity during communication, comparing the calculated transmit power with a set threshold for maximization. This research focuses on optimizing transmit power using advanced methodologies, aiming to increase secret capacity within power constraints.

Chapter 8 studies the impact of cloud computing paradigm on IoT—integrated industrial systems. This study explores the dynamic intersection of cloud computing and IIoT, examining their reflective impact on various sectors.

Chapter 9 finds the fusion of edge, IIoT, and AI, transforming Industrial Engineering and minimizing security threats in context of industrial systems integrating IoTs and other advanced technologies.

Chapter 10 gives the convergence of medical IoT (MIoT) and patient privacy, studies its challenges and solutions. It reads the prevailing issues in context of healthcare applications comprising of MIoT. It advocates for proposed solutions for cyberattacks on MIoT, giving future directions to new readers and researchers.

Chapter 11 develops an ecosystem toward a trusted smart city with the amalgamation of Internet of Everything and blockchain. It proposes a cognitive framework for shared business services and provides safe and secure transactional framework.

About the Editors

Dr. Sunil Kumar Chawla is working as Assistant Professor in the Department of Computer Science and Engineering, Chitkara University, Rajpura, Punjab, India. He has received his doctorate from IKG Punjab Technical University, Kapurthala, Punjab, India. He has more than 19 years of experience. His research interests lie in digital image processing, biometrics, image segmentation, artificial intelligence, machine learning and pattern recognition. He is a senior member of *IEEE* and other professional societies like *ISTE, CSTA, IAENG*, and *Institute of Scholars*. He has published more than 50 articles in international journals or conferences of repute, indexed in *IEEE Xplore*, Scopus, WoS, Google Scholar, and other eminent databases. He is associated with various international journals in the capacity of reviewer and guest editor. He has served as Technical Program Committee Member, Reviewer, Session Chair, and Program Chair of many conferences. He has edited two books under the publication of Routledge and CRC Press, Taylor & Francis Group, UK. He is also dealing with various academic and administrative duties for inclusive growth of the institution being served by handling teaching, learning, research, education, and overall development activities. He is also indulged in NBA, ABET, NIRF, and other accreditation and other ranking frameworks.

Dr Neha Sharma working as Assistant Professor in the Department of Computer Science and Engineering Department at Chitkara University Institute of Engineering & Technology, Chitkara University, Rajpura, Punjab, India. She has received her Ph.D. degree in the area of Computer Science Engineering from GD Goenka University, Gurugram, and has vast teaching experience of more than 14 years in a reputed organization. She has more than 80 international publications in reputed peer-reviewed journals indexed in Scopus and WoS omit including IEEE Xplore, SCOPUS, and SCI indexed. Her main area of research is the image processing, machine learning, deep learning, and cyber security. She has also published more than 40 National & International Patents under the Intellectual Property Rights of Government of India & Abroad. She is actively associated with NAAC preparations at the university interface. She is associated with many highly impact society memberships, like IEEE (senior member), ACM, and ISTE (lifetime member).

Dr. Ahmed A. Elngar is Associate Professor and Head of the Computer Science Department at the Faculty of Computers and Artificial Intelligence, Beni-Suef University, Egypt. Dr. Elngar is also, an Associate Professor of Computer Science at the College of Computer Information Technology, American University in the Emirates, United Arab Emirates. Also, Dr. AE is an Adjunct Professor at School of Technology, Woxsen University, India. Dr. AE is the Founder and Head of the Scientific Innovation Research Group (SIRG). Dr. AE is a Director of the Technological and Informatics Studies Center (TISC), Faculty of Computers and Artificial Intelligence, Beni-Suef University. Dr. AE has more than 150 scientific research papers published in prestigious international journals and over 35 books covering such diverse topics as data mining, intelligent systems, social networks, and smart environments. Dr. AE is a collaborative researcher He is a member of the Egyptian Mathematical Society (EMS) and International Rough Set Society (IRSS). His other research areas include the Internet of Things (IoT), network security, intrusion detection, machine learning, data mining, and artificial intelligence. Big Data, authentication, cryptology, healthcare systems, automation systems. He is an editor and reviewer of many international journals around the world. Dr. AE won several awards including the Young Researcher in Computer Science Engineering", from Global Outreach Education Summit and Awards 2019, on 31 January 2019 (Thursday) in Delhi, India. Also, he was awarded the Best Young Researcher Award (Male) (Below 40 years)", Global Education and Corporate Leadership Awards (GECL-2018), Plot No-8, Shivaji Park, Alwar 301001, Rajasthan.

Dr. Prasenjit Chatterjee is currently Professor of Mechanical Engineering and Dean (Research and Consultancy) at MCKV Institute of Engineering, West Bengal, India. He is also a Distinguished Researcher at the School of Engineering, Universidad Católica del Norte, Chile. He has over 6900 citations and 130 research papers in various international journals and peer- reviewed conferences. He has authored and edited more than 43 books on intelligent decision-making, fuzzy computing, supply chain management, optimization techniques, risk management, and sustainability modeling.

Prof. Chatterjee has been consecutively enlisted in the Top 2% list of researchers and scientists for a single year published by Elsevier-Scopus and Stanford University under "Operations Research" with a global rank of 3513 and a composite c-score of 2.8080. He has been ranked 9th among all Indian researchers and scientists in "Operations Research".

Dr. Chatterjee is the Editor-in-Chief of *Journal of Decision Analytics and Intelligent Computing*. He has also been the guest editor of several special issues in different SCIE/Scopus/ESCI (Clarivate Analytics) indexed journals. He is the Lead Series Editor of Disruptive Technologies and Digital Transformations for Society 5.0, Springer. He has also been the guest editor of several special issues in different SCIE/Scopus/ESCI (Clarivate Analytics) indexed journals. He is the Lead Series Editor of "Smart and Intelligent Computing in Engineering", Chapman and Hall/CRC Press, Founder and Lead Series Editor of "Concise Introductions to AI and Data Science", Scrivener-Wiley; AAP Research Notes on Optimization and Decision-Making Theories; Frontiers of Mechanical and Industrial Engineering, Apple Academic Press, co-published with CRC Press, Taylor and Francis Group and "River Publishers Series in Industrial Manufacturing and Systems Engineering". Dr. Chatterjee is one of the developers of two multiple-criteria decision-making methods called Measurement of Alternatives and Ranking according to Compromise Solution (MARCOS) and Ranking of Alternatives through Functional mapping of criterion sub-intervals into a Single Interval (RAFSI).

Dr. P Naga Srinivasu is Associate Professor in the Computer Science and Engineering Department at Amrita School of Computing, Amrita Vishwa Vidyapeetham,

AP, India. He obtained his Bachelor's degree in computer science engineering from SSIET, JNTU Kakinada (2011), and a Masters's degree in computer science technology from GITAM University, Visakhapatnam (2013). He was awarded a doctoral degree by GITAM University for his thesis on Automatic Segmentation Methods for Volumetric Estimate of damaged Areas in Astrocytoma instances Identified from the 2D Brain MR Imaging. His fields of study include biomedical imaging, soft computing, explainable AI, and healthcare informatics. He has published numerous publications in reputed peer-reviewed journals and has edited book volumes with various publishers like Springer, Elsevier, IGI Global, and Bentham Science. He was an active reviewer for more than 40 journals indexed in Scopus and Web of Science. He also served as guest editor and technical advisory board member for various internationally reputed conferences.

Contributors

Vijayalakshmi A
Department of Statistics and Data
    Science
CHRIST (Deemed to be University)
Bengaluru, India

M. Anoop
Department of Computer Science and
    Applications
SRM Institute of Science and
    Technology, Ramapuram Campus
Ramapuram Kanchipuram, India

Ansul
Department of Computer Science and
    Engineering
COER University
Roorkee, India

Shaloo Bansal
Department of Electronics &
    Communication Engineering
Mewar University
Gangarar, India

Eugene Berna I
Department of Artificial Intelligence
    and Machine Learning
Bannari Amman Institute of Technology,
Erode, India

Deepika Bhatia
Vivekananda Institute of Professional
    Studies, Technical Campus
New Delhi, India

Ashvini Chaudhari
School of CSIT
Symbiosis Skills and Professional
    University
Pune, India

Sunil Kumar Chawla
Department of Computer Science &
    Engineering
Chitkara University Institute of
    Engineering and Technology
Chitkara University
Rajpura, India

Gudditi Chetan
School of Computer Science and
    Engineering
VIT-AP
Amaravati, India

Sweta Dargad
School of CSIT
Symbiosis Skills and Professional
    University
Pune, India

Puja Das
Department of Computer Science and
    Engineering
and
Department of Computer Science and
    Engineering
Baba Farid College of Engineering &
    Technology
Bathinda, India

Venkata Lakshmi Dasari
School of Computer Science and
    Engineering
VIT-AP
Amaravati, India

Pankaj Dhiman
Department of CSE/IT
Jaypee University of Information
  Technology (JUIT)
Solan, India

Ahmed A. Elngar
Faculty of Computers and Artificial
  Intelligence
Beni-Suef University
Beni-Suef City, Egypt
and
College of Computer Information
  Technology
American University in the Emirates
United Arab Emirates

Chitra Jain
Department of Computer Science and
  Engineering
COER University,
Roorkee, India

J Jeyalakshmi
Department of Computer Science and
  Engineering
Amrita School of Computing
Amrita Vishwa Vidhyapeetham
Chennai, India.

Vijay K
Department of Computer Science and
  Engineering
Rajalakshmi Engineering College
Chennai, India

Y Lalitha Kameswari
Department of Electrical and
  Electronics Engineering
MLR Institute of Technology
Hyderabad, India

Ismail Keshta
Computer Science and Information
  Systems Department
College of Applied Sciences
AlMaarefa University
Riyadh, Saudi Arabia

Sonu Kumar
Department of Electronics and
  Communication Engineering
K.S.R.M. College of Engineering,
Kadapa, India

Kotha Lavanya
School of Computer Science and
  Engineering
VIT-AP
Amaravati, India

Yash Mane
School of CSIT
Symbiosis Skills and Professional
  University
Pune, India

Ragini Mokkapati
School of Computer Science and
  Engineering
VIT-AP
Amaravati, India

Venkatanarayana Moram
Department of Electronics and
  Communication Engineering
K.S.R.M. College of Engineering
Kadapa, India

Aruna Pavate
School of CSIT
Symbiosis Skills and Professional
  University
Pune, India
and
Department of Scientific Research,
  Innovation and Training of Scientific
  and Pedagogical Staff
University of Economy and Pedagogy
Uzbekistan

K. Raghava Rao
Department of Computer Science and
  Engineering
Koneru Lakshmaiah Education
  Foundation
Vaddeswaram, Guntur, India

Keerthi Rohan
Department of Computer Science and
    Engineering
Amrita School of Computing
Amrita Vishwa Vidhyapeetham
Chennai, India

Prithi Samuel
Department of Computational Intelligence
School of Computing
SRM Institute of Science and Technology,
    Kattankulathur Campus
Chennai, India

Sara Sawant
School of CSIT
Symbiosis Skills and Professional
    University
Pune, India

Shirley Sheeba S
Department of Computer Science
Mount Carmel College
Bengaluru, India

Sunanna S S
Department of Computer Science
Mount Carmel College
Bengaluru, India

Ashima Shahi
Department of Computer Science &
    Engineering, CGC Technical Campus
Jhanjheri, Mohali, India

Neha Sharma
Department of CSE/IT
Jaypee University of Information
    Technology (JUIT)
Solan, India

Navruzbek Shavkatov
The Department of Corporate Finance
    and Securities
Tashkent State University of
    Economics
Tashkent, Uzbekistan

Shilpi Shital
TNO (Netherlands Organization for
    Applied Scientific Research)
The Hague, Netherlands

Moutushi Singh
Department of Information
    Technology
Institute of Engineering &
    Management
Salt Lake, India

Mukesh Soni
Dr. D. Y. Patil Vidyapeeth
Pune, India
and
Dr. D. Y. Patil School of Science &
    Technology
Tathawade, India

Sarwath Unnisa
Department of Computer Science
Mount Carmel College
Bengaluru, India

Vrince Vimal
Computer Science and Engineering
Graphic Era Hill University, Graphic
    Era Deemed to be University
Dehradun, India

1 Protecting AI-Enabled Industrial Engineering in Cloud and Edge Environments

Venkata Lakshmi Dasari, Ragini Mokkapati, Kotha Lavanya, and Gudditi Chetan

1.1 INTRODUCTION

1.1.1 OVERVIEW OF IIoT AND AI IN INDUSTRIAL ENGINEERING

The field of industrial engineering is rapidly changing due to two revolutionary technologies: artificial intelligence (AI) and the Industrial Internet of Things (IIoT). The scaling up of Internet of Things (IoT) into industrial domains and applications is referred to as IIoT. In order to collect, exchange, and use data for increased productivity and decision-making, it entails networking and linking industrial machines, systems, and devices. AI is the creation of algorithms and systems that have potential to carry out tasks that normally call for human intellect. AI is used in industrial engineering to improve decision-making, automate tasks, and optimize processes.

1.1.2 UNIQUE SECURITY CHALLENGES POSED BY CLOUD AND EDGE ENVIRONMENTS

Organizations must handle the particular security issues raised by cloud and edge computing to assurance privacy, availability, and integrity of their systems and data. Some challenges are listed below:

- In order to guarantee seamless protection without sacrificing speed, meticulous planning is necessary when integrating security solutions across cloud and edge settings.
- Security risks must be continuously monitored in both cloud and edge contexts. Mechanisms for real-time detection and response are crucial.
- Migrating data and apps between cloud providers or edge solutions can be difficult for organizations, which raises the possibility of vendor lock-in.
- To reduce the risks of human error and social engineering attacks, security awareness and training for users and administrators are essential in cloud and edge systems.

DOI: 10.1201/9781003466284-1

1

In order to overcome these obstacles, a thorough and flexible security plan that takes into account the particularities of cloud and edge settings is needed. To create a strong security posture, this combines technological controls, regulations, and user education.

1.2 NETWORK ARCHITECTURE FOR IIoT

1.2.1 Overview of IIoT Network Architecture

The IIoT network architecture design is a challenging undertaking that needs careful planning to guarantee the safe, effective, and dependable operation of industrial equipment. The IIoT network architecture is shown in Figure 1.1.

An overview of the essential factors and elements of an IIoT network design is provided below:

- **Sensors and Actuators:** These are the actual hardware components that gather and send data. They must support IoT protocols like MQTT or CoAP for communication.

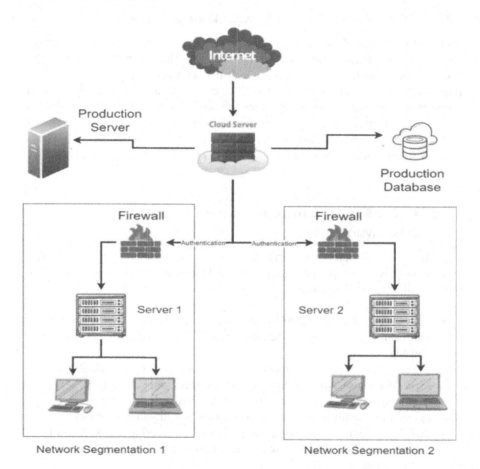

FIGURE 1.1 IIoT network architecture.

- **Edge Devices:** Equipment that preprocesses data at the network's edge before sending it to a central system or the cloud. PLCs, or Programmable Logic Controllers is an example of edge devices.
- **Communication Layer:** Based on the particular use case and environment, select the appropriate communication innovations, such as Wi-Fi, Ethernet, cellular, LPWAN (Low Power Wide Area Network), or even satellite.
- **Mesh Networks:** Mesh network deployments occasionally improve network dependability by enabling devices to relay data via other devices, hence offering redundancy.
- **Edge Computing Layer:** Reduce latency and bandwidth consumption by processing data locally by deploying computer resources at the edge. Real-time decision-making and analytics are capabilities of edge servers. To effectively manage apps at the edge, use containerization technologies like Docker or Kubernetes.
- **Gateway Layer:** They serve as go-betweens for edge devices and the core network, gathering information and managing sensor-to-cloud connectivity. Use Intrusion Detection systems and firewall rules as well as other security measures to guard against assaults on gateways.
- **Cloud or Central System:** For additional processing and storing, data from edge devices is transferred to the central system (cloud or data center). Make sure the cloud infrastructure can accommodate large size of data produced by IIoT devices.
- **Analytics and Machine Learning:** Make use of cloud resources for predictive maintenance, machine learning, and sophisticated analytics.
- **Security Layer:** For data that is in transit, use secure communication protocols and encryption (such SSL/TLS). Enforce access control procedures strictly to guarantee that devices and users may only access the network and data with permission. To monitor network activity and identify anomalous behavior, use intrusion detection and prevention systems.
- **Zero Trust Model:** Even if a user or device is trying to connect within the network, assume nothing and make sure they are authentic.

1.2.1.1 Monitoring and Management

Install tools to monitor the security, dependability, and performance of the IIoT network in real time.

- **Remote Management:** To guarantee the continuous operation and security of IIoT devices and edge components, enable remote administration and upgrades.
- **Redundancy and Reliability:** Use redundancy at different levels, such as backup systems, redundant connections, and dual power sources, to guarantee network dependability. In the event of a failure, switch to backup parts or connections by using failover techniques.
- **Regulatory Compliance:** Ascertain that the IIoT network conforms to industry-specific laws and guidelines (such as ISA/IEC 62443 in industrial environments).

- **Scalability:** Make plans for future expansion and the inclusion of additional hardware and data. Ensure that the architecture can grow with the number of IIoT deployments.

In order to build a system that satisfies the unique demands and security requirements of the industrial environment, network architects, security experts, and domain specialists should collaborate while designing an IIoT network architecture.

1.2.2 DESIGNING IIoT NETWORKS FOR SECURITY

It is essential to consider security while designing IIoT networks in order to guard sensitive data, assure the dependability of vital equipment, and stop unwanted access [1]. The following are important factors and recommended procedures for creating secure IIoT networks:

- **Risk Assessment:** For the purpose of identifying potential risks and vulnerabilities as well as their effects on the IIoT network, conduct a thorough risk assessment. This serves as the cornerstone for creating a strong security plan.
- **Network Partition:** As a result of this, intruders are prevented from moving laterally, and the impact of security breaches is reduced.
- **Secure Communication:** For data in transit, use secure transmission protocols. Data transferred between devices and systems is maintained private and safe from prying eyes, thanks to encryption [1].
- **Authentication and Authorization:** Use strong authentication strategies within the IIoT network. To further restrict access based on roles and responsibilities, create strong authorization procedures.
- **Device Management:** Use secure device management techniques, such as the capacity to revoke compromised devices, frequent firmware updates, and secure boot procedures.
- **Secure Boot and Firmware Validation:** In order to prevent tampering or unauthorized alterations, make sure that devices in the IIoT network have secure boot processes and verify the integrity of firmware and software upgrades.
- **Physical Security:** Access points, equipment rooms, and data centers must all be secured.
- **Monitoring and Logging:** Establish ongoing IIoT network monitoring to find irregularities and possible security incidents. Forensic analysis should be facilitated by the implementation of logging and auditing methods [1].
- **Update and Patch Management:** To guarantee timely deployment, provide a strong update and patch management procedure.
- **Firewalls and Intrusion Detection/Prevention Systems (IDPS):** Deploy firewalls to control traffic and IDPS to identify and respond to harmful actions within the IIoT network [2].
- **Data Integrity:** Use methods to protect the integrity of the data, including digital signatures and checksums, to identify and stop unauthorized changes [1].

- **Supply Chain Security:** Vet and protect the supply chain for IIoT components and devices to prevent the introduction of compromised hardware or software.
- **Employee Training:** To lower the danger of social engineering attacks, regularly train staff members who manage or interact with IIoT systems on security awareness.
- **Regulatory Compliance:** Assure adherence to pertinent industry standards and laws that control data security and privacy in industrial environments.
- **Redundancy and Resilience:** Redundancy and resilience should be incorporated into the IIoT network's design to assure uninterrupted executions in the case of malfunctions or assaults [3].
- **Incident Response Plan:** In case of a security incident, create an incident response plan that specifies what should be done. Test the plan periodically to ensure its efficacy.

By including these ideas in the design and continuing administration of IIoT networks, enterprises may increase the security posture of their industrial systems and limit the risk of cyber threats. To keep up with new threats and technological advancements, security measures must be updated and subject to regular security audits.

1.2.3   BEST PRACTICES FOR NETWORK SEGMENTATION AND ACCESS CONTROL

Access control and network segmentation are essential elements of a strong cyber security plan. The following are recommended procedures for dividing up a network and controlling access:

- To identify assets and their criticality, perform a detailed inventory and classification of the devices, apps, and data on your network.
- Based on the hazards associated with each asset's classification, divide the network into security zones. The demilitarized zone (DMZ), the interior, and the exterior are common zones.
- Divide the network into segments according to how the systems and devices work. To avoid cross-contamination, for instance, keep corporate IT systems and manufacturing systems apart [2].
- Implement micro-segmentation to further separate individual segments into smaller, isolated zones. This makes it possible to regulate communication between particular devices precisely.
- Create logical sections that are separate from the physical network infrastructure by utilizing network virtualization technologies. This offers scalability and flexibility [3].
- To reduce the chance of compromise and to lessen the influence that devices may have on other areas of the network, separate IoT and IIoT devices into separate segments.
- To allocate permissions according to job roles, use role-based access control (RBAC). This guarantees that users don't have unauthorized access permissions and may carry out their tasks without any extra privileges.

- Install centralized identity management systems to expedite access control, user provisioning, and de-provisioning inside the company.
- To implement access control regulations, make use of automated technologies. This promotes consistent application of security regulations and lessens the possibility of human error [4].
- To prevent unwanted access to sensitive data even in the event that perimeter defenses are compromised.
- Network firewalls should be positioned strategically to control data flow between segments and execute access control regulations.
- Create an incident response method for swiftly removing access privileges in case of a security occurrence. As a result, the possible impact of a breach is reduced.
- To evaluate the success of access control mechanisms and pinpoint areas for development, conduct routine security audits.

1.3   RISK MANAGEMENT FOR IIoT

Risk management is an important part of the IoT deployment process since the IIoT requires connecting numerous systems and devices in industrial contexts, which may introduce new vulnerabilities and potential dangers. Figure 1.2 shows the risk management in IIoT.

1.3.1   IDENTIFYING AND ASSESSING RISKS

- **Risk Identification:** List all of the IIoT systems and devices that are a part of your company. This encompasses all linked equipment and infrastructure, as well as sensors, actuators, and gateways. Assess any risks and weaknesses that can impact your IIoT infrastructure. This could involve operational hazards, physical threats, and cybersecurity threats [5]. Make sure you are aware of all applicable industry-specific laws, rules, and guidelines (such ISA/IEC 62443 and the NIST Cybersecurity Framework) that pertain to your IIoT implementations.
- **Risk Assessment:** Evaluate each risk's impact and probability that has been found. Think about things like the possibility of data breaches, business interruptions, safety risks, and monetary losses. Sort IIoT systems and assets according to how important they are to operations. Prioritize safeguarding the most important parts first. Create scenarios for various risk circumstances so you may see how they might unfold and what effects they might have [6].

1.3.2   MITIGATING RISKS THROUGH SECURITY CONTROLS

- Put best practices and security procedures in place to reduce identified risks. Regular security upgrades, intrusion detection systems, access control, and encryption are a few examples of this.
- To prevent tampering and unwanted access to IIoT devices and infrastructure, implement physical security measures.

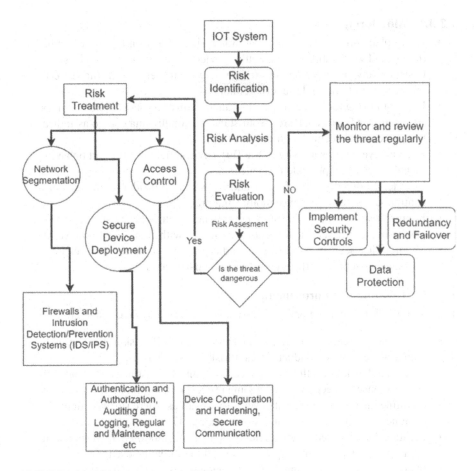

FIGURE 1.2 Risk management in IIoT.

- Make sure secure protocols and encryption are used for data transmission over secure communication channels [5].
- Divide your IIoT network into segments to separate important components from less important ones and lessen the attack surface.
- Develop an incident response strategy that states how to respond to and handle security incident recovery. Test and revise this plan often [5].
- Update all software, gateways, and IIoT devices with the most recent security fixes.

1.3.3 MONITORING AND CONTINUOUS IMPROVEMENT

For IIoT systems, effective risk management necessitates constant improvement and monitoring. By using these procedures, firms can evaluate the efficacy of current risk mitigation strategies, remain alert to emerging dangers, and adjust to shifting IIoT conditions. An outline of IIoT risk management monitoring and ongoing development is provided below.

1.3.3.1 Monitoring

- Put in place processes for ongoing monitoring to stay on top of new cyber threats and vulnerabilities that could impact IIoT systems and devices.
- Consider any irregularities or strange activity in network traffic patterns that can point to possible security concerns.
- Use monitoring tools to evaluate the functionality and state of IIoT devices. Atypical trends or departures from standard conduct may serve as indicators of potential security concerns.
- Keep an eye on user activity in the IIoT environment to spot odd behavior, illegal access, and possible insider threats.
- In order to quickly identify and address security incidents, implement a continuous incident response monitoring procedure.
- Keep an eye out for IIoT system compliance with industry standards and laws to make sure security measures are in line with necessary frameworks.
- Monitor IIoT security performance indicators, including incident response times, access control efficacy, and security patch status.

1.3.3.2 Continuous Improvement

To ensure ongoing enhancement, the subsequent actions may be considered:

- To find new threats, evaluate current risks, and modify risk mitigation techniques as necessary, conduct frequent risk assessments [6].
- Create feedback loops that allow security audits, risk assessments, and incident responses to feed into the ongoing process of continuous improvement.
- Examine the results of security events to find areas where IIoT risk management needs to be improved and lessons gained [7].
- Create adaptable security policies that may be modified in response to shifting IIoT deployment requirements and the threat landscape.
- In order to address emerging threats, vulnerabilities, and best practices in IIoT security, personnel training programs should be updated and improved [8].
- Keep up with security technology developments, and update IIoT systems to take advantage of new security features and capabilities.
- To improve IIoT security practices, interact with industry associations, exchange details about security issues, and gain insight from other companies' experiences.
- Keep up with standards and regulation changes to make sure IIoT systems continue to meet changing needs.

Organizations may increase the security posture of their industrial systems and successfully address emerging risks by using IIoT risk management practices that are based on a proactive monitoring approach and a culture of continuous improvement.

1.4 DATA SECURITY FOR IIoT

1.4.1 OVERVIEW OF IIoT DATA SECURITY

Because IIoT systems entail the gathering, transfer, and analysis of sensitive and vital data from numerous industrial devices and sensors, data security is essential in the context of the IIoT. Ensuring the integrity, confidentiality, and availability of

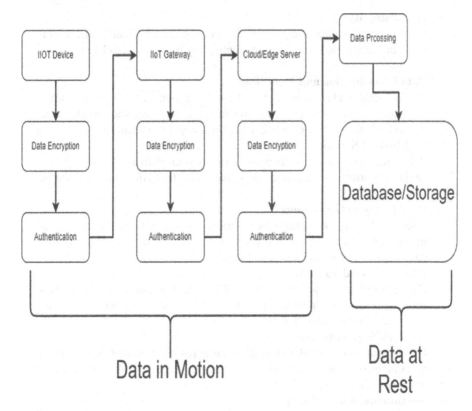

FIGURE 1.3   IIoT data security architecture.

the systems depends on protecting this data. Figure 1.3 shows the IIoT data security architecture.

These are the main IIoT data security considerations.

- **Encryption:**
  To secure data when it is being transported between devices, sensors, gateways, the central system, or the cloud, use standard encryption protocols [9].
- **Access Control:**
  Make sure that approach to IIoT devices and data is restricted to permitted users and devices only. Put strong authentication methods in place for individuals and devices [10].
- **Authorization:**
  Clearly define RBAC policies to restrict what devices and users can do inside the IIoT network.
- **Secure Protocol:**
  Make use of secure transmission protocols made for IoT applications. These protocols have security protections built right in [9].
- **Device Security:**
  Implement device-specific safety measures, such as intrude detection and secure boot, and restrict access to IIoT devices to keep them safe.

- **Data Integrity:**
  Put procedures in place to guarantee data integrity. Checksums and digital signatures are included in order to identify any unauthorized changes made to data.
- **Data Classification and Handling:**
  Assign data a classification based on legal and sensitivity standards. Implement suitable security measures in accordance with the classification. It's possible that not all data needs the same degree of security.
- **End-to-End Security:**
  Put in place end-to-end security that extends from individual devices to the cloud or central infrastructure. This guarantees that information is safe on its way [11].
- **Data Lifecycle Governance:**
  State regulations for data storage deletion, and retention in order to handle the increasing amounts of data produced by IIoT devices. Make sure that data is only kept for as long as it is required.
- **Privacy Considerations:**
  Consider privacy issues, particularly if IIoT data includes sensitive or personal data. Observe applicable data privacy laws, such as the General Data Protection Regulation (GDPR).
- **Network Segmentation:**
  Divide your IIoT network into segments to separate important systems from less important ones. This can lessen attackers' ability to move laterally and assist contain breaches.
- **Monitoring and Logging:**
  Use real-time logging and monitoring to find and look into security events. Examine logs for irregularities and possible dangers.
- **Physical Security:**
  Prevent unwanted access to the physical infrastructure of IIoT systems, such as control rooms and data centers.
- **Regulatory Compliance:**
  Make sure that your IIoT data security procedures adhere to industry-specific laws and guidelines, such as ISA/IEC 62443 in the case of an industrial setting.

To develop a safe and resilient ecosystem, data security for IIoT is a continuous process that calls for a combination of technological protections, policy, and personnel training. Ensuring the integrity of essential operations and safeguarding sensitive industrial data should be integral components of your overall IIoT strategy.

1.4.2 Securing Data in Motion and at Rest

Ensuring the confidentiality and integrity of data in IIoT contexts requires careful consideration of both data security during transmission and data security during storage. Strong data security procedures are necessary to guard private data against illegal access, interception, and manipulation. The following are the recommended practices for IIoT data security, both in motion and at rest.

1.4.2.1 Securing Data in Motion

- **Encryption:** Use standard encryption techniques to protect data when it is broadcasted between gateways, backend systems, and IIoT devices. This stops man-in-the-middle attacks and eavesdropping.
- **Virtual Private Network (VPN):** Establish private and secure communication routes with VPNs, especially when sending data over open networks. VPNs offer a secure communication tunnel and data encryption.
- **Secure Protocols:** Select encryption and authentication-rich secure communication methods. Steer clear of employing unsafe protocols that could leave data vulnerable [11].
- **Certificate Management:** To manage digital certificates used for encryption and authentication, use reliable certificate Management System. To maintain security, update and renew certificates on a periodical basis.
- **Mutual Authentication:** To confirm the identities of both communication entities, enforce mutual authentication between IIoT devices and systems. This guarantees safe data transfer between reliable parties.
- **Secure APIs:** Confirm the usage of authentication and access controls on any APIs you use for communication. Update and audit API security mechanisms on a regular basis.
- **Network Segmentation:** Divide the IIoT network into segments and create specific channels for important data flows to communicate over. This aids in preventing sensitive data from being accessed by unauthorized devices [12].

1.4.2.2 Securing Data at Rest

- **Encryption:** Encrypt data that is stored on servers, gateways, and IIoT devices to prevent undesirable access in the event of device compromise or physical theft.
- **Secure Storage Solutions:** Select safe storage options with encryption built in. Data at rest is protected by built-in encryption capabilities included in many contemporary storage systems [13].
- **Access Controls:** Install robust access controls to limit and govern people who have access to stored data.
- **Authentication Mechanisms:** Bolster the security of the authentication systems used to access stored data. Establish multi-factor authentication, mandate strong passwords, and update access credentials frequently.
- **Physical Security:** Utilize physical security measures to prevent unwanted access to storage infrastructure and IIoT devices. This is essential to stop theft or tampering.
- **Data Lifecycle Management:** Create and put into action a plan for managing the data lifecycle. This involves routinely backing up or archiving important data and safely discarding data that is no longer needed.

- **Secure Data Disposal:** Make sure that data is safely destroyed when decommissioning IIoT devices or storage systems to stop unauthorized parties from accessing any leftover data [12].
- **Logging and Auditing:** Establish auditing and logging procedures to monitor who is accessing the data. Examine logs on a periodical basis to identify any illegal activity or unauthorized access.
- **Data obfuscation:** Use data obfuscation techniques to restrict access to important information, particularly when sharing or displaying data with users who do not require full access.
- **Regular Security Audits:** To find security gaps and make sure that safeguards are working, conduct routine security audits of data storage and access restrictions.

By integrating these techniques, businesses can create a thorough IIoT data security plan that protects sensitive data while it's being transmitted and kept on systems and devices. In IIoT installations, a strong and secure data environment is maintained by routine updates, monitoring, and adherence to industry best practices.

1.4.3   BEST PRACTICES FOR ENCRYPTION AND AUTHENTICATION

Securing data in IIoT contexts requires strong authentication and encryption procedures. The following are recommended methods for IIoT data security encryption and authentication:

1.4.3.1   Encryption Best Practices

- When encrypting data both during transmission and at rest, use standard, powerful encryption methods like advanced encryption standard (AES).
- To safeguard data at every stage of its lifecycle—from the point of creation or entrance to the point of destination—enable end-to-end encryption.
- Use secure sockets layer (SSL) or transport layer security (TLS) to encrypt communication among IIoT devices, gateways, and backend servers.
- To create, store, and exchange encryption keys in a secured manner, further, in order to use a strong key management system. Update keys often to improve security.
- Use perfect forward secrecy to ensure the safety of before or upcoming sessions is not compromised in the event that one session key is compromised.
- Make sure that cryptographic keys are provisioned securely during the device's first configuration by securing the initial bootstrapping procedure of devices.
- In order to prevent unwanted access in the event of physical theft or manipulation, encrypt data kept on IIoT devices and servers.
- For safe data exchange in IIoT networks, select secure communication protocols like AMQP over TLS, CoAPs, or MQTT over TLS [12].
- To guarantee data availability and integrity even in the event of hardware failures, implement erasure coding and data redundancy in storage systems.

- Protect sensitive information by using data masking or tokenization techniques, particularly when sharing or displaying data for particular use cases.
- Make it mandatory for IIoT devices and backend systems to authenticate each other. Before transmitting data, both parties should confirm each other's identities.
- Adopt stringent credential management procedures, such as safe authentication credential updating, transfer, and storage. Steer clear of weak or default passwords.
- Using multi-factor authentication could entail integrating smart cards, biometrics, and passwords with other variables.
- To provide users and devices with particular responsibilities and permissions, utilize RBAC. Make sure that just the resources required for their responsibilities are accessible to people and devices.
- For safe authorization and authentication, take into consideration utilizing OAuth and OpenID Connect, particularly in situations where external apps must have access to IIoT resources [13].
- Ensure that authenticated devices are permitted to connect during the IIoT network onboarding process to maintain security.
- For systems and devices, use certificate-based authentication. To ensure the legitimacy of interacting entities, digital certificates must be issued and validated.
- Over the course of a device's lifetime, manage and safeguard its identity. Make ensuring that credentials or device certificates are provisioned, updated, and revoked securely when necessary.
- To spot odd login trends or possible illegal access, keep an eye on authentication events and have anomaly detection systems in place.
- For increased security and user comfort, think about adopting biometric authentication in scenarios involving user contact.

Businesses may create a solid basis for safeguarding private data, guaranteeing secure communication, and reducing the possibility of unwanted access by implementing these best practices into their IIoT data security plans. Updating encryption keys, credentials, and security rules on a regular basis is essential to sustaining a high degree of security and responding to changing threats.

1.5 COMPLIANCE FOR IIoT

Adherence to rules and regulations is essential for the safe and lawful functioning of IIoT systems, particularly in sectors where security, privacy, and safety are critical concerns.

1.5.1 Overview of Compliance Requirements for IIoT

The IIoT has different compliance standards based on the region, industry, and type of data being handled. Organizations implementing IIoT systems must take into account a number of standard regulators and compliance frameworks, some of them are listed below:

1. GDPR
2. Health Insurance Portability and Accountability Act (HIPAA)
3. National Institute of Standards and Technology (NIST) Framework
4. ISO/IEC 27001
5. ISA/IEC 62443 (formerly ISA-99)
6. North American Electric Reliability Corporation (NERC) Critical Infrastructure Protection (CIP)
7. Federal Risk and Authorization Management Program (FedRAMP)
8. California Consumer Privacy Act (CCPA)
9. Cyber Security Act (CSA)
10. Payment Card Industry Data Security Standard (PCI DSS)
11. Telecommunications Law (China)
12. Compliance with Industry-Specific Regulations

1.5.2 NIST, IEC, AND OTHER INDUSTRY STANDARDS

Industry standards, particularly those pertaining to IIoT, are vital in assisting enterprises in adhering to best practices in cyber security. An outline of important standards from NIST, IEC, and other pertinent trade associations is provided below:

1.5.2.1 NIST Standards

- **NIST Cybersecurity Framework (CSF):** It offers best practices and guidance in five key areas to help improve cybersecurity posture: identify, protect, detect, respond, and recover.
- **NIST Special Publication 800-53:** Offers a list of security measures for government agencies and information systems. It includes a broad range of privacy and security measures that can be applied to different kinds of systems, including IIoT.
- **NIST Special Publication 800-82:** Emphasizes Industrial Control System (ICS) security and offers recommendations for incorporating cybersecurity safeguards into ICS lifecycles.
- **NISTIR 8228:** Provides guidelines to enhance risk management and cyber security initiatives on the definition and classification of IoT devices.

1.5.2.2 IEC Standards

- **IEC 62443 Series:** Industrial communication networks—network and system security: consists of a set of principles for adopting security measures in industrial contexts that address the security of control systems and industrial automation.
- **IEC 62264:** Enterprise-control system integration: focuses on integrating control and enterprise systems, offering a structure for the creation and assimilation of IIoT solutions.

- **IEC 62351 Series:** Power systems management and associated information exchange—data and communications security: offers recommendations for protecting data and communications in the power industry and also addresses the security.

1.5.2.3   Other Industry Standards

- **ISO/IEC 27001:** Information security management systems (ISMS): offers a methodical approach to protecting private company data, encompassing all aspects of information security, including IIoT technologies.
- **ISO/IEC 27019:** Information security for energy utility control systems: adapts the ISO/IEC 27001 principles to the particular needs of the energy utility industry, taking control systems into account.
- **ISO/IEC 30141:** Framework for building an IoT system: describes a framework that addresses architecture, security, and privacy in the design and implementation of IoT systems.
- **ISO/IEC 21823:** Industrial automation systems and integration—unique identification for industrial components: contributes to traceability and security in industrial systems by offering standards for the distinctive identification of industrial components.
- **IoT Security Foundation's IoT Security Compliance Structure:** This structure, created by the IoT Security Foundation, offers an organized method for determining and obtaining security compliance for IoT systems.

These standards and frameworks assist organizations in laying the groundwork for secure IIoT installations. When implementing IIoT solutions, it is imperative for organizations to assess the particular requirements associated with their industry, region, and kind of data. Following these recommendations improves the overall dependability and robustness of industrial systems in addition to enhancing cyber security.

1.5.3   BEST PRACTICES FOR COMPLIANCE

Organizations must make sure they are adhering to industry norms and regulations, especially in light of IIoT. In the context of IIoT, the following are some best practices for reaching and preserving compliance:

- Keep an eye out for updates and modifications to pertinent rules and guidelines that apply to your sector and region on a regular basis [14].
- Assign people or a compliance team the task of monitoring and guaranteeing that compliance regulations are followed.
- To find any dangers and weaknesses in your IIoT systems, do regular risk assessments [14].
- To meet compliance standards right away, including security measures in the design and development stages of IIoT systems.
- Sort and manage data based on privacy and sensitivity standards.
- Adopt strict consent management procedures, particularly when handling personal information [15].

- Ensure people and appliances have the proper access rights depending on their roles, implement RBAC.
- To improve authentication security, especially when gaining access to important systems, use MFA.
- Encrypt data both while it's moving and while it's not to prevent unwanted access.
- Maintain thorough records of all activities pertaining to compliance, risk assessments, and security regulations.
- Recognize and abide by the statutory reporting standards for security incidents.
- Verify that IIoT partners and vendors abide by all applicable laws and security best practices [16].
- Employees should receive continual instruction on compliance regulations and security best practices.
- Implement security awareness initiatives to inform staff members about possible hazards and their part in upholding compliance [17].
- To evaluate security standards and regulations, conduct internal audits.
- Hire independent auditors to verify compliance and conduct external evaluations.
- Use tracking tools to watch out for security events and noncompliance with regulations on IIoT systems.
- Participate in industry forums and team up with colleagues to exchange compliance best practices and insights.
- To guarantee a thorough comprehension of regulatory obligations and legal ramifications, consult legal counsel.
- Create feedback loops from security incidents so that compliance measures are continuously enhanced.

Organizations can improve their capacity to attain and uphold compliance in the ever-changing and dynamic field of IIoT security by implementing these best practices. A robust and compliant IIoT ecosystem is facilitated by proactive risk management, frequent evaluations, and updates to security policies.

1.6 AI IN INDUSTRIAL ENGINEERING

1.6.1 OVERVIEW OF AI IN INDUSTRIAL ENGINEERING

In industrial engineering, AI is playing a bigger and more important role. It is changing how manufacturing and industrial processes are planned, run, and optimized.

1.6.1.1 Real-world Examples of AI-Enabled Industrial Engineering

AI-enabled industrial engineering has been applied in a number of industries, improving productivity, streamlining procedures, and advancing industrial settings [18]. Here are a few instances of industrial engineering using AI in the actual world:

- **Predictive Maintenance in Manufacturing:**
  Industry: Manufacturing
  Application: AI systems examine sensor and IoT device data on machinery to forecast when equipment breakdown is likely to occur. Proactive

maintenance is made possible by this, which minimizes unforeseen outages and downtime.

- **Quality Control and Defect Detection:**
  Industry: Automotive, Electronics, and Manufacturing
  Application: On production lines, computer vision and machine learning algorithms are used to check products for flaws. AI systems can accurately detect and categorize flaws, guaranteeing the quality of the final result.

- **Energy Management in Smart Buildings:**
  Industry: Building Automation
  Application: In order to improve energy use, AI algorithms examine data from sensors, HVAC systems, and other building components. This involves modifying the heating, cooling, and lighting systems in accordance with occupancy trends, meteorological predictions, and energy costs.

- **Supply Chain Optimization:**
  Industry: Logistics and Supply Chain
  Application: AI is used to improve inventory control, route optimization, and demand forecasting, among other supply chain operations. This lowers expenses, shortens wait times, and improves productivity all around.

- **Autonomous Vehicles in Warehousing:**
  Industry: Logistics and Warehousing
  Application: Drones and automated guided vehicles (AGVs) with AI capabilities move cargo through warehouses. These systems optimize routes and avoid barriers with the use of computer vision and machine learning.

- **Process Optimization in Chemical Plants:**
  Industry: Chemical Manufacturing
  Application: AI algorithms optimize chemical processes by examining data from sensors and control systems. This involves making real-time parameter adjustments to increase yield, decrease waste, and improve safety

- **Smart Grids for Energy Distribution:**
  Industry: Utilities and Energy
  Application: AI is used in smart grids to optimize and regulate the distribution of energy. By examining data from sensors and meters, machine learning algorithms can forecast demand trends, spot abnormalities, and increase grid dependability [19].

- **Robotics in Assembly Lines:**
  Industry: Manufacturing
  Application: On assembly lines, AI-powered robots perform jobs like component picking and placement, welding, and quality control. Robots can adjust to changes in jobs and goods thanks to machine learning.

- **AI-Driven Decision Support Systems:**
  Industry: Various
  Application: AI systems evaluate huge datasets to assist management in making decisions. Predicting market trends, streamlining production schedules, and spotting cost-cutting opportunities are all included in this.

- **Digital Twins in Aerospace Engineering:**
  Industry: Aerospace

Application: AI is used to produce digital twins, which are virtual copies of real assets or systems. In aircraft engineering, these twins are utilized for predictive maintenance, performance monitoring, and simulations.

- **Health and Safety Monitoring:**
  Industry: Construction and Heavy Industries
  Application: AI-enabled systems keep an eye on employee activity and spot any safety risks using cameras and sensors. This promotes adherence to safety rules and helps prevent accidents.
- **AI in Pharmaceutical Manufacturing:**
  Industry: Pharmaceutical
  Application: AI-enabled devices use cameras and sensors to monitor worker activities and identify any potential safety concerns. This encourages following safety regulations and lessens the likelihood of mishaps.

These real-world instances highlight the various uses of AI in industrial engineering and demonstrate how these tools boost output, cut expenses, and boost overall operational effectiveness in a variety of sectors.

1.6.1.2   Benefits and Challenges of AI in Industrial Engineering

Numerous advantages of AI in industrial engineering led to improvements in productivity, efficiency, and decision-making. It does, however, also come with some difficulties that must be resolved for implementation to be successful. The advantages and difficulties of AI in industrial engineering are broken down as follows:

Benefits:
- Large volumes of data can be analyzed by AI algorithms to improve operational and manufacturing processes and boost productivity.
- By anticipating equipment problems and proactively scheduling maintenance, AI systems can decrease downtime and increase the lifespan of gear [20].
- AI can improve quality control procedures by swiftly spotting product irregularities and flaws, improving the overall quality of the product.
- Better demand forecasting, inventory control, and logistics optimization are made possible by AI, which improves supply chain operations.
- Through pattern analysis, real-time configuration adjustments, and opportunity identification for energy savings, AI systems aid in the optimization of energy consumption.
- AI helps save costs in industrial operations through automating processes, maximizing resource usage, and decreasing downtime.
- By evaluating large, complicated data sets, AI offers insightful analysis that helps decision-makers make well-informed, data-driven decisions [20].
- AI makes manufacturing processes fluid and adaptive, enabling quick and customized responses to shifts in demand or product requirements.

Challenges:
- The quality and accessibility of data are critical to AI's efficacy. Data that is inconsistent or lacking can result in erroneous forecasts and choices.

- It might be difficult to integrate AI technologies with legacy systems; compatibility and infrastructure expenditures are necessary.
- When using AI in industrial settings, infrastructure construction, training, and technology procurement frequently come with hefty upfront expenditures [19].
- Professionals with the necessary skills to comprehend, apply, and oversee AI in industrial settings are hard to come by.
- It might be difficult to guarantee compatibility between diverse AI systems and gadgets, particularly when utilizing products from multiple manufacturers.
- There are new security dangers associated with increased connection and reliance on AI, such as network vulnerabilities and data breaches.
- AI use presents ethical questions, particularly in relation to algorithmic prejudice, employment displacement, and the appropriate application of technology in decision-making.
- Adopting AI technologies may present difficulties in meeting regulatory standards and obligations, particularly in highly regulated industries [21].
- A lack of knowledge about AI on the part of some stakeholders could result in resistance, skepticism, or a misalignment with business objectives.

1.7   SECURING AI-ENABLED INDUSTRIAL ENGINEERING

Protecting sensitive data, vital systems, and operational continuity all depend on the security of AI-enabled industrial engineering. Comprehensive security measures are necessary since the incorporation of AI into industrial processes may present new risks and vulnerabilities.

1.7.1   Overview of AI Security Risks in Industrial Engineering

Industrial engineering has major issues related to AI security risks that require careful management and consideration. AI integration opens up new attack vectors and potential weaknesses that adversaries might take advantage of. An outline of the main threats to AI security in the context of industrial engineering is provided below:

- Data is essential to AI, and data security is crucial. Sensitive information's privacy and confidentiality may be jeopardized by unauthorized access, data breaches, or training data manipulation.
- In adversarial attacks, input data is manipulated to trick AI models. This could result in inaccurate forecasts in industrial settings, which would affect decision-making procedures and disrupt operations.
- There may be flaws in AI models that attackers can exploit. Adversaries may initiate focused attacks to compromise the system if they identify flaws in the model architecture [22].
- Certain AI models pose a security risk because of their inability to be explicated. Understanding and interpreting AI decisions is crucial in key industrial systems to make sure they comply with safety and security regulations.

- By introducing harmful data into training sets, data poisoning causes AI models to produce inaccurate forecasts or judgments. This could have detrimental effects on quality control or manufacturing processes in industrial engineering.
- AI models may produce false predictions or be unable to adjust to changing conditions, which could result in operational inefficiencies or safety risks. This could also happen if the models are not sufficiently trained or if the training data is skewed.
- AI systems and industrial IoT devices are frequently linked. Risks of illegal access, data interception, and possible interruption of industrial processes are brought on by the growing connection [22].
- AI systems frequently depend on parts or services from other parties. A breach in the hardware or software components that make up the supply chain may result in security flaws in the system as a whole.
- Insiders who have access to AI systems could be dangerous, whether on purpose or accidentally. Vulnerabilities could be exploited by malicious insiders, and inadvertent actions could result in data breaches or system problems.

1.7.1.1 Best Practices for Securing AI-Enabled Industrial Engineering

Industrial engineering that is AI-enabled must be secured by putting in place an extensive set of best practices to handle the particular risks and difficulties that come with integrating AI into industrial systems. The following are crucial best practices for protecting industrial engineering powered by AI.

Clearly outline data governance policies that specify the gathering, storage, and use of data in accordance with privacy laws.

- When training models, use representative and diverse datasets to increase their resistance to adversarial attacks.
- Use adversarial training strategies to strengthen the model's defenses against nefarious manipulations [23].
- Update AI models with the most recent security patches and enhancements.
- When using AI, especially in situations where safety is a concern, use interpretable models or offer justifications for decisions.
- To identify and counteract attempts at data poisoning, audit and evaluate training data on a regular basis.
- Use anomaly detection techniques to find odd data patterns that might point to poisoning.
- Evaluate and control the security risks related to third-party services or components.
- Verify the security and dependability of third-party AI algorithms by conducting in-depth code reviews [23].
- To identify and stop unwanted access, use robust user authentication and ongoing monitoring.
- Inform members of the organization about security procedures and the value of reporting unusual activity.

- To avoid catastrophic failures in safety-critical applications, implement fail-safe techniques.
- Establish and follow safety procedures to reduce the danger of AI system malfunctions.
- Evaluate the security of older systems in order to find and fix vulnerabilities.
- To reduce the possible impact of security flaws in legacy systems, use isolation approaches.
- Install real-time tracking to find irregularities and possible security breaches.
- Educate staff members on security best practices, such as identifying and reporting possible security threats.
- Regularly provide phishing awareness training to lower the danger of social engineering scams.
- Put restrictions on resources to stop attacks caused by resource exhaustion.
- To equally divide AI workloads and avoid performance loss, use load balancing.
- To safeguard data sent across dispersed components, use secure communication protocols.
- To manage access permissions in dispersed systems, use RBAC.
- To find and fix vulnerabilities in industrial systems with AI capabilities, conduct frequent security audits.
- Conduct penetration testing to evaluate the system's ability to withstand actual attacks.

Collaboration between IT, operational technology (OT), security teams, and stakeholders from many departments is necessary to implement these best practices. Maintaining a secure AI-enabled industrial engineering environment requires regular evaluations, security policy updates, and awareness of new risks.

1.7.1.2 Ensuring Reliability and Safety of AI-Enabled Systems

It is important to guarantee the dependability and security of AI-enabled systems, especially in industrial engineering settings, to avoid hazards and unfavorable outcomes. Here are some essential tactics and best practices to improve AI-enabled systems' dependability and security.

- When training AI models, make use of broad and high-quality datasets to guarantee consistent performance in a range of circumstances.
- Carry out comprehensive validation and testing to find and fix any possible problems, biases, or errors in the model.
- Choose interpretable models so that interested parties may comprehend and have confidence in the decision-making process [24].
- Make use of resources and methods that shed light on the processes AI models go through to reach certain conclusions.
- To manage significant failures and guarantee system stability, put in place redundancy and fail-safe measures.
- Provide systems with the ability to shut down quickly in the event of unexpected activity.

- Use anomaly detection in conjunction with real-time monitoring to quickly spot any departures from the expected behavior of the system [24].
- Check the health of AI models and systems frequently to make sure they are operating as intended.
- Set up and follow safety procedures that specify acceptable bounds and what to do in the event when a system malfunctions or exhibits abnormalities.
- To handle possible safety accidents, create and test emergency response plans on a periodical basis.
- Provide ways for human operators to intervene and overturn AI choices as needed by incorporating human-in-the-loop mechanisms.
- Make sure there is constant oversight by human operators who are prepared to take over in an emergency.
- Put strong cyber security safeguards in place to guard AI-enabled systems against online attacks.
- Create feedback loops to gather information about incidents and system performance so that ongoing improvement is possible.
- Give AI systems the flexibility to adjust and learn from fresh information and experiences to boost efficiency and security.
- Respect applicable industry certifications and requirements for dependability and safety.
- Make sure that the relevant regulatory standards and norms are followed.
- Ensure that the AI model architecture, training data, and decision-making procedures are all thoroughly documented.
- Create traceability systems to monitor updates and modifications made to AI models over time.
- To assess potential biases, ethical issues, and societal ramifications of AI-enabled technologies, establish ethics review boards.
- Put policies in place to deal with prejudices and guarantee equity in AI decision-making.
- Make sure operators and end users receive in-depth training so they are aware of the possibilities and constraints of AI systems.
- Organize awareness campaigns to inform users of the value of following safety procedures and reporting any problems.
- To address wear and tear, update software, and guarantee the dependability of hardware components, perform routine maintenance.
- Establish procedures for routinely updating AI models, keeping in mind evolving operational needs and technological breakthroughs.
- Hire outside specialists to carry out impartial audits and validations of AI systems and models.
- To guarantee an objective evaluation of safety and dependability, look for outside validation and inspection.
- Work together with others in the industry to exchange knowledge on best practices, incidents, and lessons discovered.
- Engage in discussions, conferences, and research to know about the developments in safety and dependability.

Organizations may lay the groundwork for AI-enabled systems that are dependable, secure, and trustworthy in a range of industrial engineering applications by implementing these best practices. Retaining the integrity of AI systems over time requires constant observation, flexibility, and a dedication to moral and security concerns.

1.8 CLOUD COMPUTING AND IIoT SECURITY

When combined, cloud computing and IIoT are two revolutionary technologies that have a lot to offer about industrial operations. But this integration also brings with it new security issues.

1.8.1 OVERVIEW OF CLOUD COMPUTING FOR IIoT

An outline of how to handle security issues when utilizing cloud computing inside the IIoT framework is provided below:

- Encrypt data while it's in transit and at rest: Secure protocols should be used to encrypt data as it travels between IIoT devices and the cloud. Encrypting data kept in the cloud is also a good idea to prevent unwanted access.
- Put in place strong identity and access management (IAM) controls: These allow you to regulate user and device access to cloud services. By doing this, you can make sure that your IIoT data and services are only accessible to authorized parties [25].
- Strictly enforce robust authentication: Use MFA to confirm users' and devices' identities before granting them access to cloud services. Clearly define authorization rules to limit access to specific information and services.
- Employ secure communication protocols: To secure data in transit, ensure secure, industry-standard encryption techniques, like SSL/TLS, are used for data sent between IIoT devices and the cloud.
- Put in place data integrity checks: To ensure that data is not tampered while transmission, use digital signatures and checksums.
- Divide networks and isolate IIoT devices: By dividing the network, you may reduce the area that could be attacked and make it more difficult for intruders to move laterally within the cloud environment [26].
- Sort data according to its level of sensitivity, then provide each group with the proper security measures. It's possible that not all data needs the same degree of security.
- Recognize and abide by industry-specific laws, rules, and guidelines, such as GDPR on data privacy and ISA/IEC 62443 in industrial settings.
- Security services such as AWS IoT Core and Azure IoT Hub are provided by cloud providers and help improve the security of IIoT deployments.
- Evaluate third-party providers: In the event that you depend on outside cloud services, make sure the security protocols and procedures of the providers match your requirements [26].

- Safeguard cloud data centers: To avoid unwanted access, confirm that the physical infrastructure of the data centers operated by your cloud provider is secure.
- Assess the security protocols of your cloud service provider (CSP) and take into account the particular security solutions they provide for IIoT implementations [27].
- Educate staff members and other interested parties on potential hazards and best practices for cloud and IIoT security.

When utilizing cloud computing, IIoT security necessitates a multi-layered strategy and cooperation among OT, cloud professionals, and IT. A secure and robust IIoT ecosystem in the cloud requires regular security audits, risk management, and up-to-date maintenance on emerging threats (Table 1.1).

TABLE 1.1
Security Challenges and Best Practices for Cloud-Based IIoT Systems

Challenges	Best practice
Data Privacy and Confidentiality	These are the issues that arise when transmitting and storing critical industrial data on cloud platforms.
Network Security	Make use of virtual private networks (VPNs), IDPS, and firewalls to ensure safe connection between IIoT devices and the cloud.
Identity and Access Management (IAM)	Adopt IAM results to guarantee appropriate access controls, authorization, and authentication. Apply the least privilege concept to restrict access according to roles.
Device Security	Utilize security-by-design techniques while creating IIoT devices. To guarantee the integrity of device firmware, update and patch devices on a regular basis and utilize secure boot methods.
Data Integrity	Use integrity checks to confirm the validity and consistency of data at different points along its path from sensors to the cloud, such as digital signatures or checksums.
Security Compliance	Conduct routine compliance audits and evaluations of the cloud environment. Select cloud service providers (CSPs) who offer compliance paperwork that upholds pertinent requirements.
Data Residency and Sovereignty	Select CSPs that provide alternatives for data sovereignty and residency. Recognize and share with others the precise locations where the data will be kept.
Integration with Legacy Systems	Perform comprehensive security evaluations prior to integration. Establish secure gateways and protocols to facilitate cloud-to-legacy system connection.
Incident Response	Establish a regular incident response strategy for cloud-based IIoT systems. Use cloud-native security services to receive warnings and monitoring in real-time.
Supply Chain Security	Examine and choose reliable suppliers for cloud services and IIoT devices. Verify the security procedures of suppliers.
Data Backup and Recovery	Put frequent data recovery and backup measures into place. Make use of cloud-based backup options and do routine recovery process testing.
Shared Responsibility Model	Make sure you know exactly what your duties are as an organization and as a CSP. Apply policies and procedures within the area of the organization in accordance with the shared responsibility model.
Insider Threats	The best course of action is to use analytics and monitoring of user activity to spot odd trends. Conduct regular employee training on security awareness.

1.8.1.1 Security Challenges and Best Practices for Cloud-Based IIoT Systems

IIoT solutions that are cloud-based provide many benefits, including accessibility, scalability, and adaptability. They do, however, also provide unique security issues that should be carefully considered. These are a few typical security issues with cloud-based IIoT systems and the recommended fixes for them.

Adopting a comprehensive strategy that integrates organizational policies, technology controls, and continuous monitoring is necessary to implement a strong security framework for cloud-based IIoT systems [28]. Maintaining the security and integrity of cloud-based IIoT settings requires upgrading security procedures on a regular basis and keeping up with new threats. Reputable CSPs that give security and compliance top priority in their solutions should be selected by organizations.

1.9 EDGE COMPUTING AND IIoT SECURITY

The IIoT relies heavily on edge computing because it brings computing power nearer to the data source, increasing efficiency and lowering latency.

1.9.1 OVERVIEW OF EDGE COMPUTING FOR IIoT

In industrial settings, edge computing locates computing power near data-generating equipment including sensors, machinery, and control systems. Real-time or almost real-time decision-making is made possible by edge computing, which processes data nearer to the source, cutting down on the amount of time needed for analysis and action. By minimizing the necessity to transfer massive quantities of unprocessed data to centralized cloud servers, edge computing maximizes network capacity and lowers data transmission expenses (Table 1.2).

1.10 CASE STUDIES

1.10.1 REAL-WORLD EXAMPLES OF IIoT SECURITY AND AI-ENABLED INDUSTRIAL ENGINEERING

Here are some examples of IIoT security and AI-enabled industrial engineering in the actual world as of January 2022, when I last updated my knowledge. Remember that things have changed in this field since then, so it's best to check the most recent sources for the most recent data.

- **Siemens' Industrial Security Solutions:**

  Overview: Global industrial technology leader Siemens offers complete industrial security solutions. Their solutions include cyber security protocols for ICS and make use of AI for anomaly and threat identification.

Key Features:
AI algorithms integrated to identify anomalous network traffic patterns.

TABLE 1.2

Security Challenges and Best Practices for Edge-Based IIoT Systems

Challenges	Best Practices
Physical Security	To safeguard edge equipment, use security measures together with access controls, surveillance, and secure enclosures.
Device Security	When developing edge devices, make use of security-by-design concepts, secure boot mechanisms, and frequent firmware updates. Put device authorization and authentication into practice.
Network Security	Put in place robust network security measures, for example, secure communication protocols, firewalls, and IDPS. Keep edge networks safe from outside attacks [29].
Data Integrity:	Use cryptographic methods to confirm the integrity and validity of data processed at the edge, such as digital signatures or checksums.
Edge-to-Cloud Communication	When transmitting data between edge devices and the cloud, use safe and encrypted communication protocols (such as HTTPS). To communicate securely over public networks, use VPNs.
Identity and Access Management (IAM)	To govern and regulate access to edge devices, put IAM solutions into practice. Use robust authentication procedures, and make sure that access rights are routinely reviewed and updated.
Edge Device Authentication	For edge device access, use robust authentication methods like two-factor authentication. Turn off the default login information and change your credentials frequently [30].
Edge Analytics Security	Use secure coding techniques, frequent code audits, and secure analytics frameworks as security safeguards for edge analytics operations [30].
Data Storage at the Edge	To prevent unwanted access, encrypt data kept at the edge. Adopt safe data storage procedures and conduct frequent vulnerability audits on stored data.
Edge Device Updates	Establish a quick and safe method for updating edge devices' software and firmware. When possible, take into account over-the-air (OTA) updates [31].
Distributed Security Policies	It could be difficult to enforce uniform security guidelines among dispersed edge devices.
Edge Device Monitoring	Use anomaly detection and intrusion detection technologies to monitor edge devices continuously. Create a thorough incident response strategy for situations involving edge security [31].

Industrial networks are continuously analyzed and monitored for potential security risks.

Working together with cyber security specialists to create and execute strong security protocols.

- **Maersk and IBM's Trade Lens:**

Overview: IBM and Maersk developed Trade Lens, a blockchain-based platform for shipping. It incorporates IIoT components for safe and transparent supply chain management, even if it is not exclusively AI-focused.

Key Features:

Makes use of blockchain technology to maintain safe, unchangeable supply chain records.

Cargo tracking and monitoring is made possible in real time by IoT sensors mounted on shipping containers. Increases overall supply chain security, decreases fraud, and increases transparency.

- **General Electric's (GE) Digital Twin Technology:**

Overview: GE is well-known for its digital twin innovation, which develops virtual equivalents of tangible assets. AI is used to improve this technology in industrial settings for predictive maintenance, asset performance optimization, and overall efficiency gains.

Key Features:

AI systems use sensor data from physical assets to forecast maintenance requirements.

Proactive decision-making is made possible by the virtual, real-time representation of equipment that digital twins offer.

Prolongs the life of industrial assets, lowers downtime and enhances equipment reliability.

- **ABB's Ability™ Predictive Maintenance for Drives:**

Overview: Industrial drives can benefit from predictive maintenance programs provided by global technology giant ABB. The system tracks the condition of electric drives and anticipates possible problems using IIoT and AI.

Key Features:

IIoT sensors gather load, temperature, and vibration data from drives.

In order to forecast future failures, AI algorithms examine data patterns.

Allows for proactive maintenance, which lowers unscheduled downtime and boosts equipment dependability.

- **Bosch's IoT and AI in Manufacturing:**

Overview: Leading technology and services provider Bosch uses AI and IoT technologies to boost productivity and quality in manufacturing operations.

Key Features:

IoT sensors gather information from manufacturing processes and equipment.

AI algorithms examine data for process improvement, predictive maintenance, and quality control.

Incorporating cyber security safeguards to guarantee industrial data security.

- **Rockwell Automation's Factory Talk Analytics:**

Overview: Factory Talk Analytics is an industrial analytics platform provided by Rockwell Automation that makes use of AI to enhance overall equipment effectiveness (OEE) and optimize manufacturing operations.

Key Features:

AI algorithms examine data from networked devices to find inefficiencies and maximize output.

Utilizing predictive analytics to plan maintenance can minimize downtime.

Improved security measures to safeguard private manufacturing information.

These case studies show how top industrial businesses are optimizing many parts of industrial engineering, increasing security, and increasing efficiency by utilizing IIoT and AI. In order to handle the potential and problems in industrial settings, more businesses are anticipated to integrate new technology as the area continues to expand.

a. **Best practices from case studies**

From the case studies and best practices mentioned in the field of IIoT security and AI-enabled industrial engineering, here are some key takeaways:

- **Comprehensive Security Measures:**
  Case Reference: Siemens' Industrial Security Solutions

  Implement a comprehensive security strategy that includes cybersecurity measures for ICS. This involves a combination of AI-based threat detection, continuous monitoring, and collaboration with cybersecurity experts.

- **Blockchain for Supply Chain Security:**
  Case Reference: Maersk and IBM's TradeLens

  Utilize blockchain technology to enhance supply chain security. It gives decentralized and tamper-resistant ledger, ensuring transparency, reducing fraud, and creating a secure record of transactions in the supply chain.

- **Digital Twins for Predictive Maintenance:**
  Case Reference: GE Digital Twin Technology

  Implement this technology for assets in industrial settings. This involves creating virtual replicas of physical assets and utilizing AI algorithms to analyze real-time data from sensors. Predictive maintenance based on digital twins can significantly improve equipment reliability.

- **Predictive Maintenance for Drives:**
  Case Reference: ABB's Ability™ Predictive Maintenance for Drives

  Leverage IIoT sensors on industrial equipment, such as drives, to collect real-time data on parameters like temperature, vibration, and load. Predict potential failures with AI algorithms to prevent downtime and improve equipment efficiency.

- **IoT and AI in Manufacturing:**
  Case Reference: Bosch's IoT and AI in Manufacturing

Integrate IoT sensors into manufacturing processes to collect data on equipment and processes. Apply AI algorithms for quality control, predictive maintenance, and process optimization. Additionally, prioritize cybersecurity measures to protect sensitive manufacturing data.

- **Industrial Analytics for Process Optimization:**
  Case Reference: Rockwell Automation's FactoryTalk Analytics

  Implement industrial analytics platforms that leverage AI to analyze data from connected devices. This analysis can identify inefficiencies in manufacturing processes and contribute to optimizing production. Include security features to safeguard manufacturing data.

- **Proactive Maintenance Planning:** Emphasize proactive maintenance planning based on predictive analytics. Algorithms can analyze historical data, search for patterns, and predict equipment failures based on that data. It reduces unplanned downtime and extends asset lifespan.
- **Continuous Monitoring and Anomaly Detection:** Establish continuous monitoring mechanisms, including AI-based anomaly detection. This ensures real-time detection of abnormal patterns in network traffic, providing early indications of potential security threats.
- **User Training and Collaboration:** Best Practice: Prioritize user training and collaboration with experts in the field. Educate personnel on security best practices and ensure ongoing collaboration with cyber security experts to stay ahead of evolving threats.
- **Secure Communication Protocols:** Use of secure protocols, such as HTTPS and VPNs, to ensure the secure transmission of data between edge devices and central cloud servers. This helps protect against eavesdropping and tampering.

These best practices highlight the significance of integrating advanced innovations, such as AI and IoT, into industrial processes while concurrently addressing security considerations.

1.11   CONCLUSION

1.11.1   RECAP OF KEY CONCEPTS AND BEST PRACTICES

The way industrial operations are carried out is changing dramatically as a result of the integration of edge computing, cloud computing, and IIoT. This confluence presents previously unheard-of chances for greater productivity, better decision-making, and increased efficiency across a range of industries, from manufacturing and agriculture to healthcare and energy. But this technological development also presents a number of intricate security issues and concerns that must be taken into account. Securing this integration is essential to maximizing the potential of these technologies and protecting sensitive data and vital systems, as the case studies and talks above have shown.

Key Takeaways:

- Industries frequently follow set rules and guidelines. It is essential to make sure these guidelines are followed. For example, the GDPR for data protection and ISA/IEC 62443 in industrial contexts are crucial standards.
- In real-time data processing, edge devices are essential. To guarantee the integrity of these devices, secure boot procedures, endpoint security programs, and physical security measures must be considered.
- For security issues to be quickly identified and mitigated, thorough incident response strategies and ongoing monitoring and logging are necessary.
- In order to match security measures and best practices, it is imperative that IT and OT specialists bridge their differences. This kind of cooperation is vital in the IIoT era.
- Ensuring the integrity of edge devices and cloud data centers requires the implementation of access controls and secure enclosures.
- Programs for awareness and adequate training are essential. Stakeholders and employees alike need to be aware of the value of security and their part in risk mitigation.
- Security protocols must be customized for each sector, application, and use case. One context may not always translate well into another.

Best Practices:

- Establish a thorough security plan for IIoT systems that includes cybersecurity safeguards, ongoing observation, and expert cooperation to counter possible attacks.
- Leverage blockchain technology to improve supply chain security through transaction security, fraud reduction, transparency, and a decentralized, tamper-resistant ledger.
- Use AI algorithms for real-time analysis and decision-making, build virtual replicas of actual assets, and apply digital twin technologies for predictive maintenance.
- Place a strong emphasis on predictive analytics-based proactive maintenance planning. AI can be used to examine past data, spot trends, and forecast future equipment malfunctions.
- Install AI-based anomaly detection systems for continuous monitoring to find unusual network traffic patterns and get early warnings of any security risks.
- To ensure the secure transmission of data between edge devices and central cloud servers, defend against manipulation and eavesdropping by using secure communication protocols, such as HTTPS and VPNs.
- To identify anomalous network traffic patterns and receive early alerts of potential security threats, install AI-based anomaly detection systems for continuous monitoring.
- To gather data on machinery and operations in real-time, integrate IoT sensors into industrial processes. The basis for analytics, optimization, and decision-making is this data.

- Use AI algorithms in manufacturing for decision-making, process optimization, and quality control. This boosts productivity, lowers mistakes, and increases industrial processes as a whole.
- Utilize edge computing to operate data in IIoT systems in real-time or almost in real-time. By doing this, the source of the data generation may make decisions more quickly, while also reducing latency and optimizing bandwidth.

These concepts and best practices underscore the importance of integrating advanced technologies like AI and IoT into industrial processes while concurrently addressing security concerns. A holistic and proactive approach is key to realizing the potential benefits of IIoT and AI-enabled industrial engineering.

In conclusion, the seamless convergence of IIoT, edge computing, and cloud computing is not without its challenges, but the potential benefits in terms of efficiency, productivity, and data-driven decision-making are substantial. Furthermore, as technology evolves and new threats emerge, it is essential to stay informed and adapt security strategies accordingly to maintain a secure and resilient ecosystem.

1.11.2 FUTURE DIRECTIONS FOR IIoT SECURITY AND AI-ENABLED INDUSTRIAL ENGINEERING

IIoT security and AI-enabled industrial engineering will continue to advance due to continued technology developments, changing industry demands, and a greater emphasis on tackling new issues. In the upcoming years, the following major trends and directions are expected:

- High-tech cyber security measures will become more important as IIoT ecosystems get more complicated. This entails the application of strong encryption techniques to protect industrial data, the integration of blockchain for safe transactions, and AI-driven threat detection.
- Edge computing and AI integration will get smoother. More and more edge devices will run AI algorithms natively, allowing for real-time decision-making without primarily relying on cloud processing. Applications that need minimal latency, such process optimization and predictive maintenance, must follow this trend.
- AI-driven predictive maintenance will keep developing, thanks to increasingly complex algorithms that make use of deep learning and machine learning methods. This will result in more precise equipment failure forecasts, which will further reduce downtime and increase asset reliability.
- There will be an exponential increase in the number of IoT devices connected in industrial settings. Strong security mechanisms, such as device authentication, secure communication routes, and extensive access controls, will be required to counteract the growing attack surface as a result of this proliferation.
- IIoT security will see a rise in the use of zero trust security approaches. This method is predicated on the idea that no entity—internal or external

to the network—should be trusted by default. The core elements of IIoT security methods will be least privilege access, strict identity verification, and ongoing monitoring.

- The growing popularity of AI applications in industrial settings will lead to an increased emphasis on privacy-preserving AI. AI models will be able to learn from decentralized data sources without jeopardizing the privacy of individual user data, thanks to techniques like homomorphic encryption and federated learning.
- There will be heightened endeavors within the industry to establish standards and promote interoperability in order to guarantee the smooth integration of IIoT devices and systems. Having common standards will make it easier for companies to work together, simplify implementation, and improve system compatibility overall.
- Humans' place in industrial processes will change as a result of more AI collaboration. There will be an increase in the use of human-in-the-loop systems, in which AI enhances human decision-making. The flexibility and dependability of industrial systems are improved by this method.
- Explainable AI models will become more widely used, particularly in vital industrial applications. It is essential to comprehend how AI makes certain decisions in order to establish credibility and guarantee adherence to industry rules and guidelines.
- The attention will shift to addressing edge AI-specific security concerns as edge computing becomes increasingly integrated with IIoT. It will be crucial to guarantee edge device security, secure connectivity, and defense against edge-specific attacks.
- There will be a greater focus on AI governance, ethics, and responsible AI practices as AI becomes more and more integrated into industrial processes. To guarantee the ethical application of AI, prevent bias, and handle societal effects, organizations will implement frameworks and rules.
- Frameworks unique to IIoT and AI security will be created and improved by governments and regulatory organizations. Adopting strong security measures by firms will become essential to industrial engineering procedures as a means of complying with these rules.
- Advanced threat intelligence system development and implementation will become essential for IIoT security. In order to detect and anticipate sophisticated cyber threats, these systems will employ AI to evaluate enormous datasets, strengthening proactive cyber security measures.
- Businesses will put more and more emphasis on making their IIoT systems resilient. This entails having a thorough disaster recovery strategy, testing reaction mechanisms on a regular basis, and having the capacity to bounce back fast from physical or cyber security disturbances.
- Quantum computing presents opportunities and difficulties for IIoT security as it develops. It will be necessary for organizations to investigate encryption techniques that are resistant to quantum computing in order to safeguard their systems from any possible risks.

In summary, a concentrated effort will be made to advance technology, fortify security measures, and solve ethical and regulatory issues in the future directions of IIoT security and AI-enabled industrial engineering. The need for innovation, efficiency, and resilience in the face of a more linked and intelligent industrial world will drive the industry's continued evolution.

REFERENCES

1. Urrea, Claudio, and David Benítez. "Software-defined networking solutions, architecture and controllers for the industrial internet of things: A review." Sensors 21.19 (2021): 6585.
2. Bedhief, L. Foschini, P. Bellavista, M. Kassar, and T. Aguili, "Toward self-adaptive software defined fog networking architecture for IIoT and Industry 4.0," 2019 IEEE 24th International Workshop on Computer Aided Modeling and Design of Communication Links and Networks (CAMAD), Limassol, Cyprus, 2019.
3. Mantravadi, Soujanya, et al. "Securing IT/OT links for low power IIoT devices: Design considerations for industry 4.0." IEEE Access 8 (2020): 200305–200321.
4. Mcginthy, Jason M., and Alan J. Michaels. "Secure industrial Internet of Things critical infrastructure node design." IEEE Internet of Things Journal 6.5 (2019): 8021–8037.
5. Adaros Boye, C., Kearney, P., and Josephs, M. (2018). Cyber-Risks in the Industrial Internet of Things (IIoT): Towards a Method for Continuous Assessment. In: Chen, L., Manulis, M., Schneider, S. (eds) Information Security. ISC (2018).
6. Adaros-Boye, C., Kearney, P., Josephs, M. (2020). Continuous Risk Management for Industrial IoT: A Methodological View. In: Kallel, S., Cuppens, F., Cuppens-Boulahia, N., Hadj Kacem, A. (eds) Risks and Security of Internet and Systems. CRiSIS (2019).
7. Schurmann, J., Elchouemi, A., Prasad, P.W.C. (2023). Industrial Internet of Things Cyber Security Risk: Understanding and Managing Industrial Control System Risk in the Wake of Industry 4.0. In: Mukhopadhyay, S.C., Senanayake, S.N.A., Withana, P.C. (eds) Innovative Technologies in Intelligent Systems and Industrial Applications. CITISIA 2022.
8. Salamai, Abdullah Ali. "An integrated model for ranking risk management in Industrial Internet of Things (IIoT) system." Neutrosophic Sets and Systems 55.1 (2023): 13.
9. Yu, Yong, et al. "Toward data security in edge intelligent IIoT." IEEE Network 33.5 (2019): 20–26.
10. X. Yu and H. Guo, "A Survey on IIoT Security," *2019 IEEE VTS Asia Pacific Wireless Communications Symposium (APWCS)*, Singapore, 2019, pp. 1–5, doi: 10.1109/VTS-APWCS.2019.8851679.
11. Wang, Jin, et al. "Data security storage mechanism based on blockchain industrial Internet of Things." Computers & Industrial Engineering 164 (2022): 107903.
12. Sharghivand, Nafiseh, and Farnaz Derakhshan. "Data security and privacy in industrial IoT." AI-Enabled Threat Detection and Security Analysis for Industrial IoT (2021): 21–39.
13. Fang, Liming, et al. "A secure and fine-grained scheme for data security in industrial IoT platforms for smart city." IEEE Internet of Things Journal 7.9 (2020): 7982–7990.
14. Hassani, HichamLalaoui, et al. "Vulnerability and security risk assessment in a IIoT environment in compliance with standard IEC 62443." Procedia Computer Science 191 (2021): 33–40.
15. Bicaku, Ani, Markus Tauber, and Jerker Delsing. "Security standard compliance and continuous verification for Industrial Internet of Things." International Journal of Distributed Sensor Networks 16.6 (2020): 1550147720922731.

16. Mugarza, Imanol, Jose Luis Flores, and Jose Luis Montero. "Security issues and software updates management in the industrial internet of things (IIoT) era." Sensors 20.24 (2020): 7160.
17. Oberhofer, Daniel, Markus Hornsteiner, and Stefan Schönig. "Market Research on IIoT standard compliance monitoring providers and deriving attributes for IIoT compliance monitoring." arXiv preprint arXiv:2311.09991 (2023).
18. Zhang, Xianyu, et al. "A reference framework and overall planning of industrial artificial intelligence (I-AI) for new application scenarios." The International Journal of Advanced Manufacturing Technology 101 (2019): 2367–2389.
19. Lee, Jay, Jaskaran Singh, and Moslem Azamfar. "Industrial artificial intelligence." arXiv preprint arXiv:1908.02150 (2019).
20. Kaymakci, Can, Simon Wenninger, and Alexander Sauer. "A holistic framework for AI systems in industrial applications." Innovation Through Information Systems: Volume II: A Collection of Latest Research on Technology Issues. Springer International Publishing, 2021.
21. Rahman, MdAbdur, et al. "AI-enabled IIoT for live smart city event monitoring." IEEE Internet of Things Journal 10.4 (2021): 2872–2880.
22. Schmitt, Marc. "Securing the Digital World: Protecting smart infrastructures and digital industries with Artificial Intelligence (AI)-enabled malware and intrusion detection." Journal of Industrial Information Integration 36 (2023): 100520.
23. Rahman, Ziaur, Xun Yi, and Ibrahim Khalil. "Blockchain-based AI-enabled industry 4.0 CPS protection against advanced persistent threat." IEEE Internet of Things Journal 10.8 (2022): 6769–6778.
24. Das, Ashok Kumar, et al. "AI-envisioned blockchain-enabled signature-based key management scheme for industrial cyber–physical systems." IEEE Internet of Things Journal 9.9 (2021): 6374–6388.
25. Fu, Jun-Song, et al. "Secure data storage and searching for industrial IoT by integrating fog computing and cloud computing." IEEE Transactions on Industrial Informatics 14.10 (2018): 4519–4528.
26. Abuhasel, Khaled Ali, and Mohammad Ayoub Khan. "A secure industrial internet of things (IIoT) framework for resource management in smart manufacturing." IEEE Access 8 (2020): 117354–117364.
27. Chalapathi, G. Sai Sesha, et al. "Industrial internet of things (IIoT) applications of edge and fog computing: A review and future directions." Fog/Edge Computing for Security, Privacy, and Applications (2021): 293–325.
28. Sajid, Anam, Haider Abbas, and Kashif. "Cloud-assisted IoT-based SCADA systems security: A review of the state of the art and future challenges." IEEE Access 4 (2016): 1375–1384.
29. Wu, Yulei, Hong-Ning Dai, and Hao Wang. "Convergence of blockchain and edge computing for secure and scalable IIoT critical infrastructures in industry 4.0." IEEE Internet of Things Journal 8.4 (2020): 2300–2317.
30. Qiu, Tie, et al. "Edge computing in industrial internet of things: Architecture, advances and challenges." IEEE Communications Surveys & Tutorials 22.4 (2020): 2462–2488.
31. Hosen, ASM Sanwar, et al. "SPTM-EC: A security and privacy-preserving task management in edge computing for IIoT." IEEE Transactions on Industrial Informatics 18.9 (2021): 6330–6339.

2 Industrial Internet of Things Security Architecture and Protection Techniques

Neha Sharma and Pankaj Dhiman

2.1 INTRODUCTION

The Internet of Things (IoT) encompasses a vast and intricate network of interconnected devices, machines, and systems that are all linked to the internet. These devices are specifically engineered to simplify and enhance daily tasks by automating a wide array of processes, including but not limited to adjusting lighting, regulating thermostats, and securing entry points. As a result, they bring about increased efficiency, automation, and unparalleled convenience.

However, the security vulnerabilities in IoT devices stem from the fact that they are often designed with minimal security measures, thereby leaving them exposed to a wide range of potential cyber-attacks. These attacks can vary from relatively simple unauthorized access to a device to more advanced and sophisticated intrusions that can take control of entire networks. These security concerns underscore the importance of implementing robust security measures to safeguard IoT devices and the networks they are part of [1,2].

Updating these devices is often difficult due to the industry's complex network architecture and lack of standardization. Many IoT users are unaware of security risks associated with their devices, which lack network protection and are more vulnerable to cyber-attacks.

Hackers can exploit security weaknesses in IoT devices like Bashlite and Mirai to gain unauthorized access to sensitive data or take control of the device entirely. For example, they can use these devices to launch DDoS attacks that can cripple entire networks [3,4]. Users must know these risks and take appropriate measures to protect their devices and data. To protect IoT devices, users should follow best practices for device security, such as enabling two-factor authentication, using strong passwords, and updating device software regularly. Using a secure network and avoiding connecting to public Wi-Fi hotspots is also essential [5]. To secure IoT devices and protect private data, users should change default login credentials, keep devices updated, limit data sharing, and secure their home network. The security of IoT devices is becoming a more significant concern, and experts recommend certain measures to address them [6,7]. These include changing default passwords, updating software, and configuring devices to use secure communication protocols.

DOI: 10.1201/9781003466284-2

Recent research shows Industrial Internet of Things (IIoT) devices are prone to security risks like any other device. This puts existing IIoT installations at risk of security breaches, which can have disastrous consequences. The attacks on a German steel factory in 2014 and the Ukrainian power infrastructure in 2015 are examples of such cyberattacks. They caused power outages that affected not only a single corporation but also clients, suppliers, and even the entire country's critical infrastructure [8]. Protecting IIoT systems is critical to ensure the smooth functioning of Industry 4.0. However, implementing security measures in industrial settings can be challenging since industrial devices have a longer lifespan than consumer devices and require later security measures to ensure their continued safety [9,10]. Additionally, IIoT networks are often more extensive in scope than consumer IoT deployments, making it challenging to deploy a secure network architecture [11]. The dynamic nature of IIoT devices' increasing interconnection makes security even more challenging.

When deploying IIoT devices in Industry 4.0 settings, it is vital to comprehensively understand consumer IoT's current security flaws. This understanding will help to address known security vulnerabilities and develop suitable security measures. However, not all security techniques used in consumer IoT are relevant to the industrial world due to differences in usage, deployment, and distinct security and privacy concerns [12,13]. Therefore, specialized assessments are necessary to identify security issues and objectives in IIoT and Industry 4.0 and develop security techniques tailored to the specific characteristics of IIoT.

Research focuses on the security challenges of developing IIoT systems, including vulnerabilities, risks, and threats. A set of suitable countermeasures has been devised to address the challenges that arise from the implementation of the IIoT, including the prolonged lifespan of components and the heightened need for connectivity [14]. These measures have been developed with utmost care and expertise to ensure a smooth and efficient deployment of IIoT in various industrial settings. These measures are designed to help ensure that IIoT systems are secure, reliable, and sustainable over the long run. The work combines previously examined security techniques into a complete IIoT security survey. Our research aims to shed light on the security vulnerabilities, risks, and threats of IIoT systems. It has three key contributions:

1. We studied consumer IoT and found security changes for industrial IIoT. This study has helped to bridge the gap between consumer IoT and IIoT systems and provided a comprehensive view of the security challenges unique to IIoT.
2. We identified the security challenges faced by IIoT, which are significantly driven by safety and productivity needs. Our research provides insights into the unique security concerns the IIoT faces and offers a foundation for developing security measures that align with these key goals.
3. We examined existing research and best practices for improving IIoT security and analyzed their applicability in IIoT settings, evaluating their security benefits. This analysis has helped to identify the most effective security techniques and best practices for addressing the unique security challenges of the IIoT.

The chapter is divided into several sections. Section 2.2 provides a summary of security issues related to consumer IoT. Section 2.3 highlights the differences between security-related issues in consumer IoT and IIoT. Section 2.4 identifies unique goals and challenges faced by IIoT security. Section 2.5 presents a comprehensive survey on protecting IIoT. Also in Section 2.6, relevant work is discussed, and the article is concluded. Overall, this research provides valuable insights into the distinct security challenges faced by IIoT and lays the foundation for developing customized security measures.

2.2 THREATS TO IoT SECURITY

Research on IoT security has identified potential threats based on the architecture of the IoT. Appropriate defenses for each layer have been suggested. We have compiled similar studies on IoT security regarding attacks and countermeasures and assessed their implications for the security of the IIoT.

2.2.1 ATTACKS

IoT can be attacked in three categories: perception, network, and application. Figure 2.1 illustrates the different layers and common attacks at each level. The perception layer connects the physical environment and IoT devices. Physical attacks on IoT hardware components can compromise them [15]. The network layer connects IoT devices to the internet and can be targeted. The application layer is the foundation for IoT apps, but modification can introduce vulnerabilities.

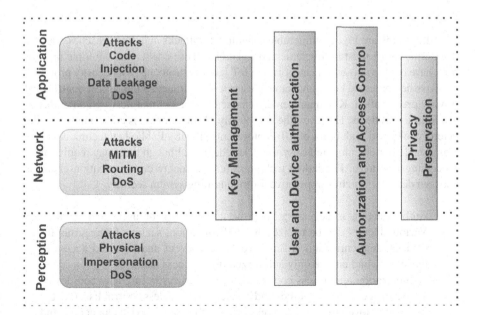

FIGURE 2.1 IoT benchmark layers include attack types and defenses.

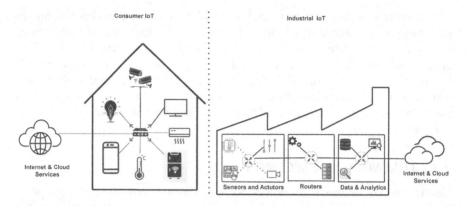

FIGURE 2.2 The IIoT (right) and consumer IoT (left) architectures. For most consumers, IoT consists of gadgets that use cloud computing to process data. Conversely, with the IIoT, cloud services are added to the local data processing to enhance workflows and offer new services.

Denial-of-service (DoS) attacks can occur at different layers. On the perception layer, they affect individual IoT devices by overloading their CPUs. On the network layer, they temporarily disconnect devices. Due to availability issues, many industrial processes that rely on this service are negatively impacted at the application layer. In conclusion, IoT vulnerabilities exist on every level, and addressing these challenges requires appropriate security measures (Figure 2.2).

2.2.2 PLANS

As IoT systems are highly vulnerable to malicious attacks and data breaches, various countermeasures must be implemented to combat these challenges. Multiple taxonomies have been proposed for categories of IoT security protocols. IoT security protocols are classified based on key management, user/device authentication, access control, and privacy preservation. Key management ensures secure communication and authentication in IoT systems. However, the complexity of this process is compounded by the heterogeneity and scalability of IoT deployments [16–18]. Public-key techniques are a manageable option compared to symmetric schemes, although they may require more processing resources. Public-key techniques create secure communication channels between devices, which is imperative for maintaining system security.

1. Authentication is necessary to prove the identity of users and devices. Various factors can be utilized for authentication, such as fingerprints for devices. Authentication is fundamental to prevent unauthorized access to the system and ensure only authorized users access data [19].
2. Authorization and access control are critical in restricting user and device access to necessary resources and services [20]. Access control lists (ACLs) can be employed to define the resources each user or device can access and their access level.

3. Privacy preservation is fundamental in safeguarding sensitive personal information, such as medical records, from unauthorized access [21]. In addition to data and communication encryption, secure multiparty computations preserve data privacy. Secure multiparty computation enables data to be processed without revealing sensitive information, ensuring privacy and confidentiality [22,23].

2.2.3 CONSIDERATION

According to the available data, while device security has been thoroughly researched and measures have been implemented to identify and mitigate potential threats, the security challenges that consumers and IIoT encounter have not been fully understood or addressed. The security concerns for consumer IoT, which focuses on the needs of individuals, differ from the security concerns for IIoT, which caters to the needs of industrial applications [24]. Developers can use the tools and approaches developed for consumer IoT, but they must also consider the unique security needs of IIoT. The usage and deployment of consumers and IIoT vary significantly, which affects the security concerns and responses required for IIoT.

IIoT devices collect and transmit private data like financial data, personal information, and intellectual property. Moreover, these devices are often installed in harsh environments like factories and power plants, where they are vulnerable to physical attacks [25,26]. Therefore, it is vital to understand IIoT security and the necessary protection mechanisms. In Section 2.5 of the report, we will provide a detailed discussion of the protection mechanisms required to ensure the security of IIoT devices.

2.3 FROM CONSUMER IoT TO IIoT

2.3.1 COMMON CONSUMER IoT AND IIoT FEATURES

IoT is a technology that connects physical devices and sensors to communicate with humans. It has two parts: consumer IoT and IIoT. Consumer IoT is designed to help people with daily tasks, such as controlling home appliances and monitoring personal health. Consumer IoT devices are typically low-cost and easy to use. They include smart speakers, fitness trackers, and home automation systems [27].

IIoT involves managing safety and mission-critical tasks in various industries, such as manufacturing, energy, and transportation. The IIoT devices monitor and control physical processes and equipment, such as pipelines, turbines, and vehicles [28]. IIoT devices use OT and IT to create M2M connections for monitoring and control. Both the consumer IoT and the IIoT have benefited from reduced costs of hardware and software, which have led to an increase in connected devices. IIoT devices are typically costly and specialized but can last up to 30 years. In contrast, consumer gadgets usually last only 3 to 5 years. Therefore, the IIoT requires scalable, adjustable, and retrofittable security solutions to protect sensitive data. The security solutions must adapt to changing threats and comply with industry standards and regulations [29,30]. The security solutions must also be integrated with the existing IT infrastructure and provide real-time monitoring and describe in Table 2.1.

TABLE 2.1
Consumer vs IIoT

Category	Characteristic	Consumer IoT	IIoT
Application	• Service Model	• Human Centered	• Machine Centered
	• Criticality	• Not Stringent	Mission Critical
Device	• Numbers of devices per home/factory lifetime	• Low to medium	• Medium to High
		• 3 to 5 years	• 10 to 30 years
	• Hardware Complexity	• Low	• Low to medium
Data Traffic	• Data Volume	• Medium	• High
	• Data Confidentiality	• Privacy-oriented	• Business–oriented
	• Traffic type	• Periodic and event-driven	• Periodic
	• Use of (wireless) communication	• Unstructured, Content-based	• Structured, Planned

Both sectors generate large amounts of data, but the IIoT mainly focuses on sensing, monitoring, and controlling duties, resulting in regular and deterministic data flows. This makes establishing network controls and detecting intrusions easier [31]. To maintain the highest standards of professionalism and responsibility, the industrial sector must take all necessary measures to safeguard the confidentiality of company secrets and protect the sensitive data of valued customers. Similarly, user privacy protection is paramount in the consumer sector and must be treated with the utmost care and diligence [32].

In conclusion, while the consumer and IIoT share similarities, they cater to different needs and demands. As such, it is essential to implement appropriate security measures to safeguard sensitive data and ensure the smooth operation of these technologies.

2.3.2 SUMMARY

IoT is a vast network of devices, ranging from consumer to industrial devices. While consumer IoT devices and IIoT devices share some technological similarities, there are significant differences in their use and implementation.

One key difference is the criticality of applications. IIoT devices are used for critical applications such as manufacturing, transportation, healthcare, and energy production, where a malfunction or a security breach can have severe consequences [33]. In contrast, consumer IoT devices are used for non-critical applications such as home automation, entertainment, and fitness tracking [34].

Another key difference is the estimated lifetime of devices. IIoT devices are designed to last for years, even decades, and are often integrated into expensive and complex systems. In contrast, consumer IoT devices have a much shorter lifespan and are often replaced after a few years [35].

Data traffic periodicity determinism is also an essential distinction between consumer and IIoT devices. IIoT devices require a higher level of determinism, meaning that the periodicity of data traffic must be consistent [36]. In contrast, consumer IoT devices can tolerate some degree of variability in data traffic periodicity.

Due to these differences, IIoT devices are more complex and challenging to secure than consumer IoT devices. While security approaches for consumer IoT devices primarily focus on individual security, the same is inadequate for IIoT devices. Security for IIoT devices must consider the interconnectivity of the devices, the criticality of the applications they serve, and the reliability of the data they generate [37,38].

Thus, security measures for IIoT devices must be entirely reevaluated. Section 2.4 will delve deeper into the security objectives and challenges arising from these differences.

2.4 IIoT SECURITY OBJECTIVES AND CHALLENGES

IIoT is now integral to several industries, including manufacturing, healthcare, and transportation. However, with increased connectivity comes a higher risk of cyber-attacks, which can cause significant financial damage, damage claims, and jeopardize human safety. To ensure the safety and productivity of IIoT, it is essential to identify potential attackers and establish unique security objectives that prioritize availability and integrity. Protecting IT infrastructure from data theft and industrial espionage is increasingly essential. Implementing IIoT security objectives in industrial environments poses significant challenges but must be addressed to prevent intrusion and sabotage [39,40].

2.4.1 ATTACKER TYPE

Accurate attacker models are necessary to protect IIoT systems from cyber-attacks. They simulate how cybercriminals would breach security and help experts understand potential risks [41]. This helps organizations comply with regulations and standards and develop effective countermeasures. Unlike other domains, IIoT requires a realistic and resilient attacker model to account for attackers engaging in malicious activities without restrictions. The Dolev–Yao model is a theoretical construct used to analyze the security of computer networks [42]. It is based on the assumption that an attacker can intercept and modify every communication in the IIoT network. This means that the attacker can eavesdrop on any message sent between two devices and change its contents before reaching its intended recipient [43]. The Dolev–Yao model is often used to study the security of cryptographic protocols and is an essential tool in designing and evaluating secure communication systems. Hence, the model enables both passive and active attacks. However, it is crucial to note that this model assumes the attacker cannot break the cryptographic techniques used in IIoT security [44].

Furthermore, it is critical to differentiate between inside and outside attackers in IIoT security. Inside attackers have more capabilities than outside attackers as they have access to IIoT devices, which enables them to launch more sophisticated attacks [45,46]. On the other hand, outside attackers can only attack IIoT devices through the network. Therefore, it is necessary to have a robust attacker model that can account for both inside and outside attackers and help identify and mitigate security risks in IIoT systems.

2.4.2   Challenges

IIoT is a network of interconnected devices used in industrial settings and designed to perform specific functions. Securing IIoT devices is a complex task that differs significantly from securing consumer IoT devices. Several factors must be considered regarding the security of IIoT devices [47].

One of the primary challenges in securing IIoT devices is the long lifespan of their components. Since IoT devices designed for industrial use have a longer lifespan than consumer devices, it is essential to consider application and communication security when developing them. Regular software updates are also necessary after deployment to maintain security. However, it can be challenging to ensure security when previously deployed devices lack security safeguards and have a complicated upgrade process. IIoT's increased connectivity raises security risks, especially when older isolated components are connected to the network [48,49].

Another challenge for the security of IIoT devices is the rising number of resource-constrained devices that require deployment, configuration, and management. Due to the vast number of devices, scalable, autonomous systems are needed to operate, create, and configure security measures [50]. These systems must be designed to manage the security of many devices, which can be a significant challenge.

The IIoT's main advantage is the vital communication between IT, OT, and the internet, allowing for more efficient and adaptive industrial production. However, dividing and isolating devices based on functionality becomes increasingly challenging, limiting unauthorized access. Network segmentation, as recommended by NIST, can protect ICSs. It limits damage caused by attacks and prevents unauthorized access. Consider it for IIoT security [51]. The IIoT collects vast amounts of data, including customer information and corporate secrets, which must be kept secure from unauthorized access. Maintaining data confidentiality while allowing authorized IIoT services access for processing and analysis is a fundamental challenge [52].

The above issues emphasize the need for innovative and customized security techniques for the IIoT. Traditional consumer IoT device techniques are unsuitable for the wide range of uses and deployments in the industrial realm. Therefore, in Section 2.5, we provide security methods for the IIoT and evaluate their effectiveness in addressing the above-mentioned concerns [53]. These methods include encryption, authentication, access control, and intrusion detection. Implementing a multi-layered approach to security is essential to protect IIoT devices from various threats.

2.5   SECURITY OF THE IIoT

We have researched security options for the IIoT to address its significant security challenges and achieve overall security goals. Our research focuses on how technologies work. The original development of security techniques for consumer IoT has led to their adaptation for use in an industrial context [54]. To facilitate understanding, we have carefully categorized the various security techniques based on broad security measures that correspond to the structure of this section. We have gone a step further to provide a comprehensive summary of the significant contributions of each strategy and an extensive overview of our discussion in Table 2.2. To help gauge

TABLE 2.2
IIoT Security Measures Comparison

	Long-lived Components	A Large Number of Devices	High Connectivity	Critical Processes	Data Confidentiality	Human Error and Sabotage
Tailored cryptography and authentication	◐	◐	○	●	●	○
Lightweight authentication and encryption [44,45]	◐	●	◐	○	●	○
Certificateless encryption [46,47]						
Patch management	◐	●	○	○	◐	○
Automated updates [48]	○	◐	○	◐	◐	◐
Detection of Vulnerabilities [49,50]						
Service Isolation and access control	○	◐	◐	●	●	◐
Trusted execution environments [51,52]	●	◐	●	○	◐	●
Network Security and Policies [53–56]						
Network monitoring and intrusion detection	●	◐	○	◐	◐	○
Network-based monitoring	●	◐	◐	●	◐	◐
Process-aware IDS [60–63]						
Awareness training and assessment	○	○	○	◐	◐	○
Awareness and training [64–66]	○	◐	○	◐	◐	◐
Assessment [67,68]						

the effectiveness of each strategy in addressing the challenges outlined in Section 2.4, we have rated each strategy based on its level of consideration, which ranges from no consideration to partial consideration to full consideration. With this in mind, we hope to provide a detailed and valuable resource for implementing security techniques in consumer IoT and industrial contexts.

2.5.1 CUSTOMIZED CRYPTOGRAPHY AND AUTHENTICATION

Encryption is vital for data confidentiality and authentication in IIoT. Resource-constrained IIoT devices require lightweight symmetric-key encryption techniques. However, symmetric-key cryptography presents challenges with key management and can cause delays for critical procedures, which may discourage industrial operators from using encryption and authentication [55].

To properly solve security concerns in IIoT, it is necessary to employ cryptography and authentication techniques customized for the IIoT. Techniques are proposed for enabling lightweight authentication and encryption in industrial communication settings, using resource-constrained devices while minimizing latency [56]. One proposed technique, as described by the proponents of [44], assumes that communication in the IoT primarily involves periodic contact with static communication partners. As such, symmetric encryption and authentication can be partially precomputed using templates for automated encryption and data authentication. Depending on the packet size, this reduces security processing by up to 76%.

For resource-constrained devices, authentication techniques that rely solely on hash and XOR operations have been suggested to provide lightweight authentication [57]. These techniques have the advantage of being computationally efficient and require minimal memory. However, such techniques require devices to be equipped with a secure element, which may not always be feasible with outdated equipment [58].

As IIoT data exchange and cloud service usage increase, protecting data from unauthorized access is essential. Some methods only secure communication with external entities, like cloud services [59,60]. One such system, [61], builds on [62] and reduces computational expenses by approximately 30%. These systems use certificates searchable by public-key encryption, facilitating simple key management across various devices. The basic concept is to secure data by encrypting it before transmitting it to the cloud service. The encrypted data can be searched, but individual data bits will be encrypted only when they are retrieved from the cloud [63]. These techniques are instrumental when relying on third-party cloud services, as outsourced information security cannot be guaranteed when dealing with a growing number of devices.

2.5.2 PATCH ADMINISTRATION

Many manufacturers are increasingly concerned about IoT devices' security and privacy concerns, as most fail to provide security updates or require significant manual effort to install patches. This leaves customers with known vulnerabilities that hackers can exploit, highlighting the importance of timely patching of all deployed IoT devices to ensure the secure operation of IIoT systems [64,65]. This not only mitigates vulnerabilities but also reduces the risk of attacks.

To fortify the security of IIoT systems, firms must improve their internal methods for promptly patching vulnerabilities, and manufacturers must regularly come up with security patches for devices throughout their operational lifetime. This is particularly important for devices deployed in industrial settings, as they may not be easily accessible for manual updates [66].

It must be kept in mind that the patching process for industrial systems often requires a lengthy testing phase before installation to ensure compatibility with the existing configuration [67]. NIST recommends regression testing for safer and more efficient patch management.

IETF proposes a mechanism for automatic firmware updates on IoT devices. The mechanism includes a standardized description of entities involved, security threats, assumptions, and secure end-to-end communication of updated device firmware [68]. This enables easy adaptation of IoT devices, and operators can deploy only those that comply with the standard. In this way, the proposed mechanism can help to increase the efficiency, safety, and security of IoT devices and networks, making them more reliable and user-friendly.

Two methods are available to detect security issues in IIoT: checking idle devices and analyzing vulnerabilities using an IIoT network graph. These approaches are valuable for safety- and mission-critical activities and serve as a good starting point for detecting current security problems and their implications for other systems [69]. Viable solutions can be designed, such as isolating affected devices to prevent further damage.

2.5.3 INTRUSION DETECTION AND NETWORK MONITORING

Implementing intrusion detection mechanisms has become crucial in identifying and preventing malicious actions that may lead to significant damage. However, deploying traditional IT networks' intrusion detection systems (IDSs) in the industrial domain is challenging due to the domination of real-time operations and resource-constrained devices in industrial control systems (ICSs) [70,71].

The intricacy of industrial networks demands the deployment of multiple IDS monitors for comprehensive traffic coverage. Despite these challenges, adopting IDSs in IIoT presents unique opportunities. The deterministic nature of industrial processes results in more regular network traffic patterns, making it easier to identify anomalies than unpredictable communication patterns in IT networks [72].

To detect unauthorized access in the IIoT, researchers use real-time, passive techniques that don't interfere with industrial processes. Inspection costs have to be reduced for scalability with growing number of devices [73]. To monitor traffic in the IoT, we can monitor it at the flow level instead of analyzing every packet. This can reduce monitored traffic by an order of magnitude, solving scalability issues in the IIoT.

Analyze processes and network behavior to improve intrusion detection in IIoT. Create a model of industrial processes using domain expertise or machine learning. Comprehensive methodologies like these have the advantage of significantly reducing detection error rates [74]. For instance, the evaluation results of this method demonstrate an achievable accuracy of 99.8%, with assaults separated from system problems. Such methodologies can also detect configuration errors operators commit to some extent [75].

2.6 RELATED WORK

IIoT is a new and innovative system that connects industrial devices, machines, and systems to the internet. Although it provides many benefits, such as increased efficiency, productivity, and cost savings, it poses significant security risks that must be addressed. The security issues in IIoT can be classified into three categories: difficulties and requirements, growing connectivity, and the security impact of industrial processes.

2.6.1 LITERATURE STUDIES ON IIoT SECURITY

Many researchers have been concerned about security issues related to IIoT, and numerous surveys have been conducted to investigate and understand these issues. These surveys cover a broad spectrum of IIoT security aspects, such as the challenges and requirements of securing IIoT systems and the security implications of industrial processes. Table 2.2 summarizes the survey findings for a comprehensive overview of the current IIoT security. Our work aims to complement and extend the existing research on IIoT security by conducting a detailed analysis of the security challenges related to IIoT and proposing practical solutions to mitigate them. Our primary objective is to enhance the security of IIoT systems and enable their widespread adoption in the industry.

Multiple studies have considered the unique security requirements and issues of IIoT, considering its components' features and interdependencies. The surveys that rely primarily on structured or systematic literature reviews, such as [76] and [77], are incredibly useful for comprehensively understanding the IIoT security research ecosystem. However, these surveys often fail to address how well the potential countermeasures fit the needs and address the corresponding security concerns.

To address this gap, our work uses the outlined security challenges and requirements described in Section 2.4. We then propose potential countermeasures and explore how well they fulfill the various criteria and security challenges outlined in Section 2.5. This approach enables us to provide a more detailed and practical analysis of IIoT security, paving the way for more effective security measures and widespread adoption of IIoT in the industry.

2.6.2 INCREASED IIoT CONNECTIVITY

IIoT has facilitated increased connectivity in ICS, which has created a significant challenge in terms of security. A recent study has examined this challenge and emphasized the need for confidentiality in sharing data across different organizations and cloud services, as mentioned in Section 2.4. Although two previous works, [78] and [79], have explored security concerns and countermeasures of cloud-based ICSs and issues related to the exchange of production data, they have not directly addressed the main challenge manufacturers face regarding local network security [80]. The study identified the need for countermeasures against inside attackers and their potential impact on local networks in the IIoT. In Sections 2.5 and 2.5.3, the study proposed adequate countermeasures to protect against inside attackers and secure local networks in the IIoT.

2.6.3 INDUSTRIAL PROCESSES' IMPACT ON SECURITY

The security of IIoT is an evolving research area that investigates the impact of industrial processes on security. This field encompasses the relationship between safety and security in IIoT, which has been extensively examined in several research papers, such as [81–82]. It posits that ensuring safety in industrial processes necessitates implementing adequate security measures. However, in some cases, security measures can contradict safety measures. Therefore, it is imperative to identify IIoT security challenges based on the conventional product life cycle, from design to maintenance, as the paper recommends [83–84].

This methodology emphasizes the importance of evaluating individual difficulties and countermeasures at each stage of the product life cycle to guarantee optimal security and safety in industrial processes.

2.7 CONCLUSION

IIoT security is crucial and requires diverse countermeasures to protect against potential threats. Customized security protocols are needed for resource-constrained devices and industrial processes, while network-layer security techniques such as access restrictions and monitoring are essential for maintaining a secure environment. Additionally, addressing insider threats through staff awareness and regular training is necessary for IIoT's security.

Exploring novel techniques to enhance standard security measures while developing IIoT is necessary. Distributed ledger technology and blockchain enable decentralized accountability in a variety of settings. Smart contracts are a promising blockchain application for enforcing access control in industrial environments. It is crucial to consider using these technologies for new device generations while considering current long-lived installations with older hardware.

The study on IIoT security is structured as follows: Section 2.2 summarizes consumer IoT security issues and countermeasures. Section 2.3 of the article compares consumer security with IIoT security and highlights the security goals and challenges of IIoT in Section 2.4. Section 2.5 presents a comprehensive survey on how to protect IIoT, while Section 2.6 discusses relevant research. The chapter concludes in Section 2.7. Security and privacy issues with consumer IoT devices have been extensively studied in recent years, and a common approach is to identify threats based on the IoT's architectural layers and implement appropriate defenses for each tier. This approach is also used in similar studies on IoT security regarding attacks and countermeasures, with their implications evaluated for IIoT security.

In conclusion, this chapter examines ways to secure the IIoT during the transition to Industry 4.0. The security challenges and goals for IIoT differ from those of consumer IoT. To address these challenges, various countermeasures are required, such as customizing security protocols for resource-constrained devices and critical industrial processes and implementing network-layer security techniques such as access restrictions and monitoring. Addressing insider threats through staff awareness and regular training is also essential.

More research is necessary to develop IIoT while enhancing standard security measures, as IIoT and Industry 4.0 are still relatively new. One way to accomplish this is to explore distributed ledger technology and blockchains, such as smart contracts, which can provide immutable and decentralized accountability in various situations. However, when employing these technologies for new device generations, it is also necessary to consider the older hardware of current, long-lived installations.

REFERENCES

1. L. Atzori, A. Iera, and G. Morabito, "The Internet of Things: A survey," *Comput. Netw.*, vol. 54, no. 15, pp. 2787–2805, Oct. 2010.
2. L. D. Xu, W. He, and S. Li, "Internet of Things in industries: A survey," *IEEE Trans. Ind. Informat.*, vol. 10, no. 4, pp. 2233–2243, Nov. 2014.
3. M. Wollschlaeger, T. Sauter, and J. Jasperneite, "The future of industrial communication: Automation networks in the IoT and Industry 4.0 era," *IEEE Ind. Electron. Mag.*, vol. 11, no. 1, pp. 17–27, Mar. 2017.
4. R. Mahmoud, T. Yousuf, F. Aloul, and I. Zualkernan, "Internet of Things (IoT) security: Current status, challenges and prospective measures," in *Proc. IEEE 10th Int. Conf. Internet Technol. Secured Trans.*, Dec. 2015, pp. 336–341.
5. Y. Yang, L. Wu, G. Yin, L. Li, and H. Zhao, "A survey on security and privacy issues in IoT," *IEEE Internet Things J.*, vol. 4, no. 5, pp. 1250–1258, Oct. 2017.
6. M. Capellupo *et al.*, "Security and attack vector analysis of IoT devices," in *Proc. Secure, Privacy, Anonymity Comp.*, Comm., Storage, 2017, pp. 593–606.
7. O. Alrawi, C. Lever, M. Antonakakis, and F. Monrose, "SoK: Security evaluation of home-based IoT deployments," in *Proc. IEEE Symp. Secure Privacy*, May 2019, pp. 1362–1380.
8. T. Yu *et al.*, "Handling a trillion (unfixable) flaws on a billion devices: Rethinking network security for the Internet-of-Things," in *Proc. ACM Workshop Hot Topics Netw.*, Nov. 2015, pp. 1–7.
9. U. Lindqvist and P. G. Neumann, "The future of the Internet of Things," *Commun. ACM*, vol. 60, no. 2, pp. 26–30, 2017.
10. A. Marzano *et al.*, "The evolution of Bashlite and Mirai IoT botnets," in Proc. IEEE Symp. Comput. Commun., Jun. 2018, pp. 813–818.
11. M. Antonakakis *et al.*, "Understanding the Mirai botnet," in *Proc.USENIX Secur. Symp.* Vancouver, BC, Canada: USENIX Association, Aug. 2017, pp. 1093–1110.
12. J. Wurm *et al.*, "Security analysis on consumer and industrial IoT devices," in *Proc. ASP-DAC*, Jan. 2016, pp. 519–524.
13. A. Sadeghi, C. Wachsmann, and M. Waidner, "Security and privacy challenges in IIoT," in *Proc. ACM/EDAC/IEEE Des. Autom. Conf.*, Jun. 2015, pp. 1–6.
14. E. Sisinni, A. Saifullah, S. Han, U. Jennehag, and M. Gidlund, "Industrial internet of things: Challenges, opportunities, and directions," *IEEE Trans. Ind. Informat.*, vol. 14, no. 11, pp. 4724–4734, Nov. 2018.
15. R. M. Lee, M. J. Assante, and T. Conway, "German Steel Mill Cyber Attack," *Ind. Control Syst.*, vol. 30, 2014, Art. no. 22.
16. D. E. Whitehead, K. Owens, D. Gammel, and J. Smith, "Ukraine cyberinduced power outage: Analysis and practical mitigation strategies," in *Proc. IEEE 70th Annu. Conf. Protective Relay Eng.*, Apr. 2017, pp. 1–8.
17. S. A. Kumar, T. Vealey, and H. Srivastava, "Security in Internet of Things: Challenges, solutions and future directions," in *Proc. Hawaii Int. Conf. Syst. Sci.*, Jan. 2016, pp. 5772–5781.
18. O. El Mouaatamid, M. Lahmer, and M. Belkasmi, "Internet of Things security: Layered classification of attacks and possible countermeasures," *Electron. J. Inf. Technol.*, vol. 9, no. 9, pp. 24–37, 2016.
19. J. Deogirikar and A. Vidhate, "Security attacks in IoT: A survey," in *IEEE Int. Conf. Social, Mobile, Analytics Cloud*, Feb. 2017, pp. 32–37.
20. M. Frustaci, P. Pace, G. Aloi, and G. Fortino, "Evaluating critical security issues of the IoT world: Present and future challenges," *IEEE Int. Things J.*, vol. 5, no. 4, pp. 2483–2495, Aug. 2018.
21. G. Vishwakarma and W. Lee, "Exploiting JTAG and its mitigation in IoT: A survey," *Future Int.*, vol. 10, no. 12, 2018, Art. no. 121.

22. J. Zhang, Z. Wang, Z. Yang, and Q. Zhang, "Proximity-based IoT device authentication," in *Proc. IEEE Conf. Comput. Commun.*, May 2017, pp. 1–9.
23. K. T. Nguyen, M. Laurent, and N. Oualha, "Survey on secure communication protocols for the Internet of Things," *Ad HocNetw.*, vol. 32, pp. 17–31, 2015.
24. M. Frustaci, P. Pace, and G. Aloi, "Securing the IoT world: Issues and perspectives," in *Proc. IEEE Conf. Standards Commun. Netw.*, Sep. 2017, pp. 246–251.
25. A. K. Das, S. Zeadally, and D. He, "Taxonomy and analysis of security protocols for Internet of Things," *Future Gener. Comput. Syst.*, vol. 89, pp. 110–125, 2018.
26. V. Oleshchuk, "Internet of Things and privacy preserving technologies," in *Proc. Int. Conf. Wireless Commun., Veh. Technol., Inf. Theory Aerosp. Electron. Syst. Technol.*, May 2009, pp. 336–340.
27. V. Hassija, V. Chamola, V. Saxena, D. Jain, P. Goyal, and B. Sikdar, "A survey on IoT security: Application areas, security threats, and solution architectures," *IEEE Access*, vol. 7, pp. 82 721–82 743, 2019.
28. N. Neshenko, E. Bou-Harb, J. Crichigno, G. Kaddoum, and N. Ghani, "Demystifying IoT security: An exhaustive survey on IoT vulnerabilities and a first empirical look on internet-scale IoT exploitations," *IEEE Commun. Surv. Tut.*, vol. 21, no. 3, pp. 2702–2733, Jul.–Sep. 2019.
29. M. A. Khan and K. Salah, "IoT security: Review, blockchain solutions, and open challenges," *Future Gener. Comput. Syst.*, vol. 82, pp. 395–411, 2018.
30. B. Galloway and G. P. Hancke, "Introduction to industrial control networks," *IEEE Comm. Surv. Tut.*, vol. 15, no. 2, pp. 860–880, Jul. 2013.
31. M. Cheminod, L. Durante, and A. Valenzano, "Review of security issues in industrial networks," *IEEE Trans. Ind. Informat.*, vol. 9, no. 1, pp. 277–293, Feb. 2013.
32. K. Stouffer *et al.*, "Guide to industrial control systems (ICS) security," NIST, Gaithersburg, MD, USA, Rep. 800–82 Rev 2, 2011.
33. H. P. Breivold and K. Sandström, "IoT for industrial automation–Challenges and technical solutions," in *Proc. IEEE Int. Conf. Data Sci. Data Intensive Syst.*, Dec. 2015, pp. 532–539.
34. M. Marjani *et al.*, "Big IoT data analytics: Architecture, opportunities, and open research challenges," *IEEE Access*, vol. 5, pp. 5247–5261, 2017.
35. D. Dzung, M. Naedele, T. P. Von Hoff, and M. Crevatin, "Security for ICS," *Proc. IEEE*, vol. 93, no. 6, pp. 1152–1177, Jun. 2005.
36. S. Adepu and A. Mathur, "Generalized attacker and attack models for cyber-physical systems," in *Proc. IEEE 40th Annu. Comput. Softw. Appl. Conf.*, 2016, pp. 283–292.
37. M. D. Ryan, "Enhanced certificate transparency and end-to-end encrypted mail," in *Proc. Netw. Distrib. Syst. Secur. Symp.*, 2014, pp. 1–14.
38. D. Dolev and A. Yao, "On the security of public key protocols," *IEEE Trans. Inf. Theory*, vol. 29, no. 2, pp. 198–208, Mar. 1983.
39. P. Priller, A. Aldrian, and T. Ebner, "Case study: From legacy to connectivity migrating industrial devices into the world of smart services," in *Proc. IEEE Emerg. Technol. Factory Autom.*, Sep. 2014, pp. 1–8.
40. J. Li, F.R. Yu, G. Deng, C. Luo, Z. Ming, and Q. Yan, "Industrial Internet: A survey on the enabling technologies, applications, and challenges," *IEEE Comm. Surv. Tut.*, vol. 19, no. 3, pp. 1504–1526, Apr. 2017.
41. A. Sajid, H. Abbas, and K. Saleem, "Cloud-assisted IoT-based SCADA systems security: A review of the state of the art and future challenges," *IEEE Access*, vol. 4, pp. 1375–1384, 2016.
42. J. Pennekamp *et al.*, "Dataflow challenges in an internet of production: A security & privacy perspective," in *Proc. ACM Workshop Cyber-Physical Syst. Secur. Privacy*, Oct. 2019, pp. 27–38.

43. J. Hiller, M. Henze, M. Serror, E. Wagner, J. N. Richter, and K. Wehrle, "Secure low latency communication for constrained industrial IoT scenarios," in *Proc. IEEE 43rd Conf. Local Comput. Netw.*, Oct. 2018, pp. 614–622.

44. A. Esfahani *et al.*, "A lightweight authentication mechanism for M2M communications in industrial IoT environment," *IEEE Internet Things J.*, vol. 6, no. 1, pp. 288–296, Feb. 2019.

45. M. Ma, D. He, N. Kumar, K. R. Choo, and J. Chen, "Certificateless searchable public key encryption scheme for industrial Internet of Things," *IEEE Trans. Ind. Informat.*, vol. 14, no. 2, pp. 759–767, Feb. 2018.

46. J. Fu, Y. Liu, H. Chao, B. K. Bhargava, and Z. Zhang, "Secure data storage and searching for industrial IoT by integrating fog computing and cloud computing," *IEEE Trans. Ind. Informat.*, vol. 14, no. 10, pp. 4519–4528, Oct. 2018.

47. B. Moran, M. Meriac, H. Tschofenig, and D. Brown, "A firmware update architecture for IoT devices," IETF, Fremont, CA, USA, Tech. Rep., drafted- suit-architecture-08, 2019.

48. J. L. Flores and I. Mugarza, "Runtime vulnerability discovery as a service on IIoT systems," in *Proc. IEEE 23rd Int. Conf. Emerg. Technol. Factory Autom.*, Sep. 2018, pp. 948–955.

49. G. George and S. M. Thampi, "A graph-based security framework for securing industrial IoT networks from vulnerability exploitations," *IEEE Access*, vol. 6, pp. 43 586–43 601, 2018.

50. C. Lesjak, D. Hein, and J. Winter, "Hardware-security technologies for industrial IoT: Trustzone and security controller," in *Proc. IEEE 41st Annu. Conf. Ind. Electron. Soc.*, Nov. 2015, pp. 2589–2595.

51. S. Pinto, T. Gomes, J. Pereira, J. Cabral, and A. Tavares, "IIoTEED: An enhanced, trusted execution environment for industrial IoT edge devices," *IEEE Internet Comput.*, vol. 21, no. 1, pp. 40–47, Jan. 2017.

52. E. Lear and B. Weis, "Slinging MUD: Manufacturer usage descriptions: How the network can protect things," *Proc. IEEE Int. Conf. Sel. Topics Mobile Wireless Netw.*, Apr. 2016, pp. 1–6.

53. A. Hamza *et al.*, "Clear as MUD: Generating, validating and applying IoT behavioural profiles," in *Proc. Workshop IoT Secu. Privacy*, Aug. 2018, pp. 8–14.

54. K. Gai, K.-K.-R. Choo, M. Qiu, and L. Zhu, "Privacy-preserving content-oriented wireless communication in Internet-of-Things," *IEEE Internet Things J.*, vol. 5, no. 4, pp. 30593067, Aug. 2018.

55. D. Henneke, L. Wisniewski, and J. Jasperneite, "Analysis of realising a future industrial network using software-defined networking (SDN)," in *Proc. IEEE World Conf. Factory Commun. Syst.*, May 2016, pp. 1–4.

56. N. Sharma and P. Dhiman, "Privacy in smart homes with remote user authenticated key establishment protocol," in *Procedia Comput. Sci.*, vol. 233, pp. 119–128, Jan. 2024, https://doi.org/10.1016/j.procs.2024.03.201.

57. M. Mantere, I. Uusitalo, M. Sailio, and S. Noponen, "Challenges of machine learning based monitoring for industrial control system networks," in *Proc. IEEE 26th Int. Conf. Adv. Inf. Netw. Appl. Workshops*, Mar. 2012, pp. 968–972.

58. T. Cruz et al., "Improving network security monitoring for industrial control systems," in *Proc. IEEE Int. Symp. Integr. Netw. Manage.*, May 2015, pp. 878–881.

59. A. Sivanathan, D. Sherratt, H. H. Gharakheili, V. Sivaraman, and A. Vishwanath, "Low-cost flow-based security solutions for smart-home IoT devices," in *Proc. IEEE Int. Conf. Adv. Netw. Telecommun. Syst.*, Nov., 2016, pp. 1–6.

60. N. Goldenberg and A. Wool, "Accurate modeling of Modbus/TCP for intrusion detection in SCADA systems," *Int. J. Crit. Infrastruct. Protect.*, vol. 6, no. 2, pp. 63–75, 2013.

61. D. Hadžiosmanoviʹc, R. Sommer, E. Zambon, and P. H. Hartel, "Through the eye of the PLC: Semantic security monitoring for industrial processes," in Proc. 30th Annu. Comput. Secur. Appl. Conf., Dec. 2014, pp. 126–135.

62. C. Zhou et al., "Design and analysis of multimodel-based anomaly intrusion detection systems in industrial process automation," IEEE Trans. Syst. Man Cybern.: Syst., vol. 45, no. 10, pp. 1345–1360, Oct. 2015.

63. J. J. Chromik, A. Remke, and B. R. Haverkort, "Bro in SCADA: Dynamic intrusion detection policies based on a system model," in Proc. Ind. Control Syst. Cyber Secur. Res., Aug. 2018, pp. 112–121.

64. U. Franke and J. Brynielsson, "Cyber situational awareness–A systematic review of the literature," Comp. Secur., vol. 46, pp. 18–31, 2014.

65. A. P. Mathur and N. O. Tippenhauer, "SWaT: A water treatment testbed for research and training on ICS security," in Int. Workshop Cyber-Phys. Syst. Smart Water Netw., Apr. 2016, pp. 31–36.

66. D. Antonioli et al., "Gamifying ICS security training and research: Design, implementation, and results of S3," in Proc. Workshop Cyber-Phys. Syst. Secur. Privacy, Nov. 2017, pp. 93–102.

67. N. Sharma and P. Dhiman, "Lightweight privacy preserving scheme for IoT based smart home," Recent Advances in Electrical & Electronic Engineering (Formerly Recent Patents on Electrical & Electronic Engineering), 16, Oct. 2023. https://doi.org/10.2174/0123520965267339230928061410.

68. H. Boyes, B. Hallaq, J. Cunningham, and T. Watson, "The industrial Internet of Things (IIoT): An analysis framework," Comput. Ind., vol. 101, pp. 1–12, 2018.

69. D. Chen et al., "Over the air provisioning of industrial wireless devices using elliptic curve cryptography," in Proc. IEEE Int. Conf. Comput. Sci. Autom. Eng., vol. 2, Jun. 2011, pp. 594–600.

70. P. Yanguo, C. Jiangtao, P. Changgen, and Y. Zuobin, "Certificateless public key encryption with keyword search," China Commun., vol. 11, no. 11, pp. 100–113, Nov. 2014.

71. M. Patton, E. Gross, R. Chinn, S. Forbis, L. Walker, and H. Chen, "Uninvited Connections: A study of vulnerable devices on the Internet of Things (IoT)," in Proc. IEEE Joint Intell. Secur. Inform. Conf., Sep. 2014, pp. 232–235.

72. E. Bertino, K.-K. R. Choo, D. Georgakopolous, and S. Nepal, "Internet of Things (IoT): Smart and secure service delivery," ACM Trans. Internet Technol., vol. 16, no. 4, pp. 22:1–22:7, Dec. 2016.

73. M. Höst, J. Sönnerup, M. Hell, and T. Olsson, "Industrial practices in security vulnerability management for IoT systems–An interview study," in Proc. Conf. Softw. Eng. Res. Pract., Aug. 2018, pp. 61–67.

74. S. Woodhouse, "Information security: End user behavior and corporate culture," in Proc. IEEE 7th Int. Conf. Comput. Inf. Technol., Oct. 2007, pp. 767–774.

75. N., Sharma and P. Dhiman, "Secure authentication scheme for IoT enabled smart homes," in Lecture notes in Electrical Engineering, 2024, Mar. 2024, pp. 611–624. https://doi.org/10.1007/978-981-99-8646-0_48.

76. M. Bada, A.M. Sasse, and J. R. C. Nurse, "Cyber security awareness campaigns: Why do they fail to change behaviour?" 2019, arXiv:1901.02672.

77. K. Tange, M. De Donno, X. Fafoutis, and N. Dragoni, "Towards a systematic survey of industrial IoT security requirements: Research method and quantitative analysis," in Proc. Workshop Fog Comput. IoT, Apr. 2019, pp. 56–63.

78. M. Lezzi, M. Lazoi, and A. Corallo, "Cybersecurity for Industry 4.0 in the current literature: A reference framework," Comput. Ind., vol. 103, pp. 97–110, Dec. 2018.

79. X. Yu and H. Guo, "A survey on IIoT Security," in Proc. IEEE VTS Asia Pacific Wireless Commun. Symp., Aug. 2019, pp. 1–5.

80. D. E. Kouicem, A. Bouabdallah, and H. Lakhlef, "Internet of Things security: A top-down survey," Comput. Netw., vol. 141, pp. 199–221, 2018.
81. S. R. Chhetri, N. Rashid, S. Faezi, and M. A. A. Faruque, "Security trends and advances in manufacturing systems in the era of Industry 4.0," in Proc. IEEE/ACM Int. Conf. Comput.-Aided Des., Nov. 2017, pp. 1039–1046.
82. Y. Pan, J. White, D. C. Schmidt, and A. Elhabashy, "Taxonomies for reasoning about cyber-physical attacks in IoT-based manufacturing systems," Int. J Interact. Multimedia Artif. Intell., vol. 4, no. 3, pp. 45–54, Mar. 2017.
83. M. N. Bhuiyan, M. M. Rahman, M. M. Billah, and D. Saha, "Internet of things (IoT): A review of its enabling technologies in healthcare applications, standards protocols, security, and market opportunities," IEEE Internet Things J., vol. 8, no. 13, pp. 10474–10498 (1 July 2021). https://doi.org/10.1109/jiot.2021.3062630.
84. Y. Zhang, S. Kasahara, Y. Shen, X. Jiang, and J. Wan, "Smart contract-based access control for the Internet of Things," IEEE Internet Things J., vol. 6, no. 2, pp. 1594–1605, 15 June 2018.

3 Risk Management in IIoT

*Sonu Kumar, Y. Lalitha Kameswari, K. Raghava Rao,
Venkatanarayana Moram, and Shilpi Shital*

3.1 INTRODUCTION

The advent of the Industrial Internet of Things (IIoT) has ushered in unprecedented opportunities for enhanced connectivity and efficiency in industrial processes. However, with this integration of digital technologies comes a complex landscape of risks that necessitate comprehensive management strategies. This chapter delves into the dynamic realm of risk management in IIoT, exploring the multifaceted challenges posed by the interconnected nature of industrial systems. In recent years, the literature has witnessed a surge in discussions surrounding cybersecurity concerns, vulnerabilities, and privacy issues within the realm of IIoT. The common risks involved in IIoT projects and applications are represented in Figure 3.1. Researchers have contributed valuable insights, presenting frameworks and models for assessing and mitigating risks specific to IIoT environments. Notably, studies such as those by Lee et al. [1] and Zakaria et al. [2] have probed into Internet of Things (IoT) cybersecurity technologies and proposed risk management models.

FIGURE 3.1 Common risks involved in IIoT projects and applications [3].

DOI: 10.1201/9781003466284-3

IIoT introduces unique challenges in risk management due to its interconnected and dynamic nature. A comprehensive review reveals key insights such as: (a) Continuous risk management methodology: Research emphasizes adapting traditional risk management standards for continuous monitoring of risk factors in IIoT networks [4]. (b) Security risk management reference model: A proposed reference model integrates mixed methods for IoT security risk management, offering a strategic approach [5]. (c) Cybersecurity risk assessment in IIoT: Addressing cyber risks, particularly through continuous risk assessment, is crucial for safeguarding IIoT networks [6]. (d) IIoT-enabled technologies and infrastructure: A study explores IIoT-enabled technologies, highlighting challenges and issues that need attention in risk management [7]. (e) Systematic review on cybersecurity risk analysis in IoT: A comprehensive analysis of 40 articles emphasizes privacy issues and cybercrimes, guiding future research directions [8]. (f) Impact of Industry 4.0 on risk management programs: Industry 4.0's transformative digitalization plays a vital role in enhancing risk management programs in IIoT.

Cybersecurity-related concerns are the most concerning when it comes to managing IoT risk, and cybersecurity is thought to be essential for businesses. For example, among the main drawbacks and shortcomings of the modern IoT are security, trust, privacy, and identity management, according to IEC [9]. The main risk impact areas of IoT governance for organizations, as determined by a survey of 374 global IoT stakeholders, are privacy and trust, safety, and security, according to World Economic Forum research [10] on the health of the IoT globally. The McKinsey & Company survey, which gathered responses from 1161 global IoT practitioners, revealed that cybersecurity was the top concern for organizations when purchasing IoT equipment, further highlighting the seriousness of the issue [11]. IoT security is the CEO's (chief executive officer's) top priority, according to AT&T [12], which made this claim in the AT&T cybersecurity insights report. Results from the McKinsey & Company study [13] show that 75% of the 400 IoT specialists asked thought that IoT security was either critical or very important. According to Bain's [14] findings, organizations place a greater emphasis on their IoT risks the more mature their cybersecurity capabilities.

Because there is no global IoT security standard [10], no general IoT security model [9], very few IoT security standards [13], and most best practices do not address IoT security risk management [1,15], IoT security risk management may therefore cause the greatest amount of concern among organizations. Furthermore, in the transformation process, cybersecurity measures are typically designed reactively as opposed to proactively [10,16,17]. Given this, it is quite likely that many organizations do not have proper IoT security risk management plans in place even though they are implementing IoT. For example, Lee [1] outlined the results of a recent survey indicating that a relatively small percentage of respondents had a cybersecurity plan in place that takes IoT security requirements into account. McKinsey & Company also noted that for a sizable majority of organizations, there are no cybersecurity plans that address IoT [13].

This chapter aims to consolidate and analyze the existing body of knowledge, providing a comprehensive overview of the various facets of risk management in IIoT. By synthesizing findings from diverse sources, the review intends to offer a holistic understanding of the challenges posed by cyber threats, vulnerabilities, and privacy concerns in the industrial landscape. Through this exploration, we seek to contribute to the development of effective risk management strategies tailored to the intricacies of the IIoT.

3.2 VARIOUS FACETS OF RISK MANAGEMENT IN IIoT

The various facets of risk management in IIoT are listed and explained in this section in detail, which are important for risk management in IIoT.

3.2.1 CONTINUOUS RISK MONITORING IN IIoT NETWORKS

Continuous risk monitoring in IIoT networks is a critical practice aimed at ensuring the ongoing security and integrity of interconnected industrial systems. This approach involves the continuous discovery, analysis, and monitoring of all operational technology, information technology (IT), and IIoT assets within the operational environment. The process incorporates incident and event management tools along with misbehavior detection mechanisms to address cyber-physical systems' security gaps. By continuously monitoring network activities and implementing predictive maintenance strategies, organizations can safeguard IIoT networks from cyber threats and unauthorized access, contributing to the overall resilience and security of industrial operations. Figure 3.2 represents a framework for continuous risk management in IIoT.

FIGURE 3.2 A framework for continuous risk management in IIoT [6].

3.2.2 Security Risk Management Frameworks for IIoT

Security risk management frameworks play a pivotal role in safeguarding IIoT eco-systems, addressing the unique challenges posed by interconnected industrial systems. The industrial internet security framework (IISF) is one such comprehensive framework that provides architectures and best practices for constructing trustworthy systems in the IIoT landscape. These frameworks encompass various risk management activities, including risk assessments, vulnerability assessments, penetration testing, and backup and recovery processes, ensuring a holistic approach to cybersecurity in the IIoT environment. Additionally, researchers have proposed innovative approaches such as the IoT security risk management strategy reference model (IoTSRM2), relying on selected IoT security best practices to enhance risk management strategies in IIoT. These frameworks contribute to the establishment of a robust security posture, addressing the evolving threat landscape in the industrial sector. Figure 3.3 represents security challenges in IIoT, and Figure 3.4 represents risk management frameworks for IIoT.

3.2.3 Cybersecurity Threats in Industrial IoT

The integration of IIoT into industrial processes brings forth various cybersecurity threats that demand vigilant attention. A systematic review of several articles emphasizes the prevalence of privacy issues and cybercrimes in the realm of IIoT, underscoring the need for robust cybersecurity measures. Figure 3.5 represents two important cybersecurity concerns.

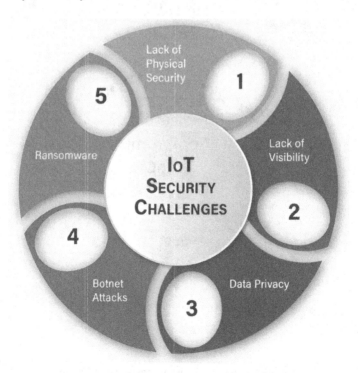

FIGURE 3.3 IoT security challenges [18].

Assessing security risks
Implementing & maintaining secure framwork
Reviewing & improving
Monitoring and recording

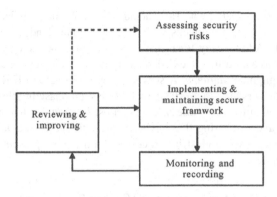

FIGURE 3.4 Security risk management frameworks for IIoT [19].

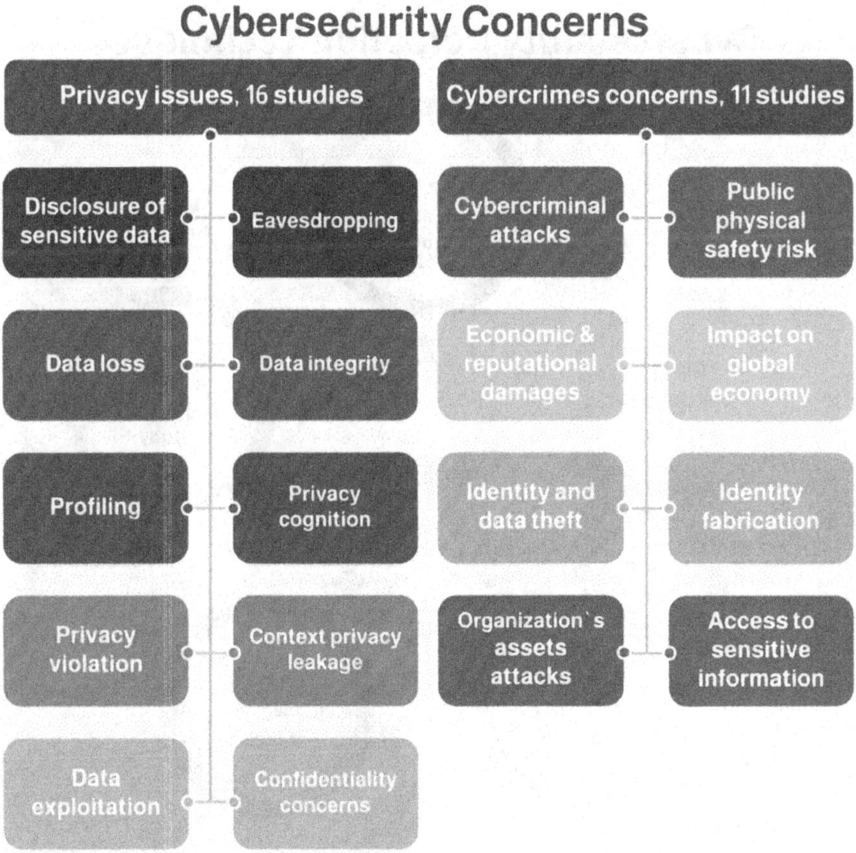

Cybersecurity Concerns
Privacy issues, 16 studies
Cybercrimes concerns, 11 studies
Disclosure of sensitive data
Eavesdropping
Cybercriminal attacks
Public physical safety risk
Data loss
Data integrity
Economic & reputational damages
Impact on global economy
Profiling
Privacy cognition
Identity and data theft
Identity fabrication
Privacy violation
Context privacy leakage
Organization`s assets attacks
Access to sensitive information
Data exploitation
Confidentiality concerns

FIGURE 3.5 Important cybersecurity concerns [8].

Common risks associated with IIoT include data siphoning, device hijacking, and distributed denial-of-service attacks, presenting a complex landscape for cybersecurity professionals. Additionally, the lack of authentication and security in process sensors is identified as a hard-to-solve IIoT security issue, posing operational challenges and potential equipment damage. A survey on cyber threats to IIoT highlights the diverse ways attackers target industrial applications, emphasizing the importance of understanding and mitigating these risks. As industries continue to embrace Industry 4.0, addressing and mitigating these cybersecurity threats is crucial to ensure the resilience and integrity of industrial processes. Figure 3.6 represents cybersecurity detection techniques.

3.2.4 DYNAMIC RISK ASSESSMENTS FOR MODERN IoT SYSTEMS

Dynamic risk assessments (DRAs) play a crucial role in ensuring the security of modern IoT systems. As highlighted in literature on security risk management for IoT, the dynamic nature of IoT environments demands adaptive risk assessment methodologies. This is particularly pertinent in the context of medical-based IoT

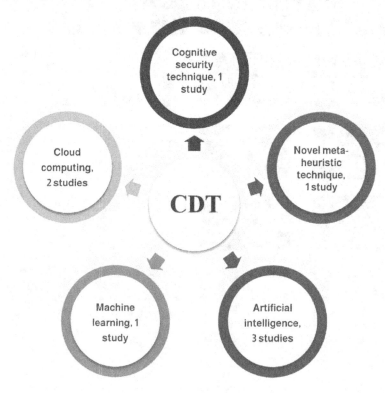

FIGURE 3.6 Cybersecurity detection techniques [8].

systems, where the challenges and opportunities for conducting DRAs are explored to establish effective protections. The need for specialized methodologies that consider the unique dynamics of IoT systems is emphasized, ensuring the rigor of best practices while addressing the evolving threat landscape. Proposals for methodologies, such as one that reconstructs time-dependent functions for optimal cybersecurity response strategies, demonstrate the commitment to real-time risk management in the face of ever-changing threats. Overall, DRAs prove essential in proactively managing and responding to risks in the rapidly evolving and interconnected landscape of modern IoT systems.

3.2.5 POLICY MANAGEMENT ESSENTIALS IN IoT RISK MITIGATION

Effective policy management is a cornerstone in mitigating the security risks associated with IoT devices. Policies serve as the foundation for preventing unauthorized access, usage, and administration of IoT devices by individuals and processes. Ensuring robust policy frameworks involves defining and implementing protocols that govern physical and logical access to these devices. This not only safeguards against potential threats but also establishes a systematic approach to manage and reduce vulnerabilities within IoT ecosystems. Recognizing the significance of policy management, organizations are advised to incorporate it as an essential element in their overall IoT security strategy. By doing so, they can proactively address potential risks and bolster the resilience of their IoT environments.

3.2.6 INTEROPERABILITY RISKS IN IIoT DEPLOYMENTS

Interoperability, or the seamless interaction between diverse devices and systems, is crucial for the success of IIoT deployments. However, it introduces inherent risks. Coexistence and interoperability challenges may arise when verifying that IoT devices or wireless networks operate accurately within a vast IIoT ecosystem. Lack of standardized communication protocols and incompatible technologies can lead to data inconsistencies and system failures, posing significant threats to the reliability and functionality of IIoT deployments. Organizations must address interoperability risks by adopting industry standards, ensuring compatibility among devices, and implementing robust testing and validation processes. By mitigating these risks, they can enhance the efficiency and effectiveness of their IIoT initiatives. Figure 3.7 summarizes interoperability taxonomy in IIoT deployments.

3.2.7 RETROFITTING CHALLENGES IN IoT RISK MANAGEMENT

Retrofitting in IoT poses unique challenges in the realm of risk management. The process of integrating IoT technologies into existing infrastructures, especially in the context of retrofitting, introduces complexities related to interoperability, security, and legacy system adaptation. One significant challenge lies in ensuring seamless communication and compatibility between the new IoT devices and the pre-existing infrastructure. Security concerns emerge as retrofitting may expose vulnerabilities in older systems, potentially compromising the integrity of the entire network.

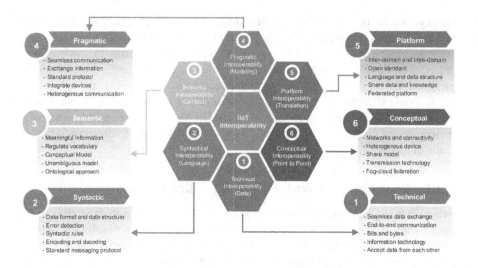

FIGURE 3.7   Interoperability taxonomy in IIoT deployments [20].

Additionally, adapting legacy equipment to modern IoT standards requires careful consideration to prevent disruptions and ensure a smooth transition. Addressing these retrofitting challenges is paramount to effective IoT risk management, requiring a strategic approach that encompasses thorough testing, security assessments, and strategic planning.

3.2.8   Real-time Monitoring Solutions for IIoT Risk

Real-time monitoring solutions play a pivotal role in mitigating risks associated with IIoT implementations. These solutions leverage advanced technologies such as smart sensors and IoT devices to provide continuous and instantaneous insights into industrial processes. By offering real-time visibility, organizations can detect anomalies, potential failures, or security breaches promptly, enabling proactive interventions. Real-time condition monitoring, as seen in applications like heat exchangers, facilitates quick response to equipment health issues. Predictive maintenance strategies, supported by network monitoring in IIoT, empower industries to address concerns before they escalate. The top IIoT monitoring solutions incorporate smart sensors and technologies, ensuring a comprehensive approach to risk management in industrial settings. As IIoT continues to revolutionize industries, real-time monitoring emerges as a critical component for maintaining operational integrity and minimizing potential risks.

3.2.9   Predictive Analytics in IIoT Risk Assessment

Predictive analytics plays a transformative role in IIoT risk assessment, leveraging past data to foresee potential future scenarios. By analyzing historical information through advanced algorithms, predictive analytics enables organizations to identify patterns, anticipate failures, and suggest proactive risk mitigation strategies. This approach not only enhances the overall security of IIoT systems but also allows for more efficient resource allocation and maintenance planning. From smart meters to predictive analytics, the integration of these technologies empowers industries to make informed decisions, reduce operational risks, and achieve sustainability goals. The use of predictive analytics in IIoT exemplifies a forward-looking approach, ensuring a proactive stance in managing risks associated with industrial processes.

3.2.10 Enhancing Security in IIoT Through Risk Management

Enhancing security in IIoT environments necessitates a proactive approach through effective risk management strategies. IIoT security challenges encompass various aspects, including administrative and operational risks, technical vulnerabilities, and potential unauthorized access. By adopting risk management practices, organizations can identify, assess, and mitigate these risks, ensuring the robustness of their IIoT systems. Strategies such as embracing a Zero-Trust Architecture and utilizing cloud-based IIoT security solutions contribute to reducing legal consequences and enhancing overall resilience. The convergence of risk management principles with IIoT security measures establishes a foundation for a secure and resilient industrial ecosystem.

3.2.11 Cost-effective Risk Mitigation in IIoT

Cost-effective risk mitigation in IIoT involves implementing strategies that prioritize resource efficiency while ensuring robust security. One approach is to conduct thorough risk assessments to identify potential threats and vulnerabilities specific to the IIoT environment. Leveraging proven risk mitigation strategies such as avoidance, reduction, transference, and acceptance can be tailored to align with cost-effectiveness. For example, avoiding unnecessary risks, reducing the impact through preventive measures, transferring certain risks through insurance or partnerships, and accepting low-impact risks can collectively contribute to a balanced and economical risk management plan. This proactive cost-effective approach helps organizations maintain the integrity of their IIoT systems without unnecessary financial burdens.

3.2.12 Role of AI in Industrial IoT Risk Management

Artificial intelligence (AI) plays a pivotal role in IIoT risk management by leveraging advanced analytics and machine learning techniques. AI enhances the security of IIoT systems by identifying and mitigating potential threats and vulnerabilities in real-time. Figure 3.8 represents the convergence of AI and IoT.

Its application extends to risk analytics, where AI and machine learning models are employed to analyze data in IIoT networks, particularly in industrial settings, aiding in effective risk assessment and management. Additionally, AI-based applications in IIoT contribute to avoiding unplanned downtime, introducing new products and services, and enhancing overall risk management strategies. The use of AI techniques, including machine learning and deep learning, is explored for cybersecurity management in Industry 4.0, showcasing the potential for intelligent risk mitigation.

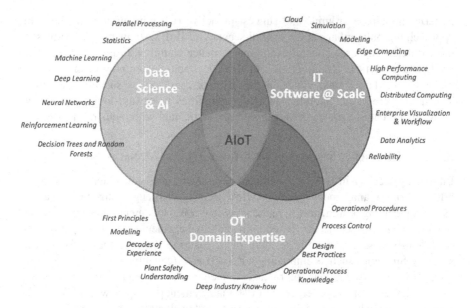

Parallel Processing
Statistics
Machine Learning
Deep Learning
Neural Networks
Reinforcement Learning
Decision Trees and Random
Forests

Data Science & AI

Cloud
Simulation
Modeling
Edge Computing
High Performance
Computing
Distributed Computing
Enterprise Visualization
& Workflow
Data Analytics
Reliability

IT Software @ Scale

AIoT

First Principles
Modeling
Decades of
Experience
Plant Safety
Understanding
Deep Industry Know-how

OT Domain Expertise

Operational Procedures
Process Control
Design
Best Practices
Operational Process
Knowledge

FIGURE 3.8   Convergence of AI and IoT [21].

This intersection of AI and IIoT is not only vital for identifying and managing risks but also for ensuring the robustness and resilience of industrial systems.

3.2.13   EDGE COMPUTING IMPACT ON IIoT RISK

Edge computing has a profound impact on IIoT risk dynamics. While it enhances IIoT capabilities by processing data closer to the source, minimizing latency, and reducing the need for centralized processing, it introduces unique security challenges. The proximity of edge devices to the physical infrastructure increases the risk of unauthorized access and potential breaches. Security concerns include edge security risks that, if not properly addressed, can compromise smooth business operations and lead to data loss or theft. Understanding and mitigating these risks become critical to ensuring the secure implementation of IIoT systems leveraging edge computing. Strategies for overcoming edge security risks involve comprehensive measures to protect against unauthorized access and potential vulnerabilities in the distributed architecture of edge computing within IIoT environments. Figure 3.9 represents balance between edge and cloud computing in IIoT.

3.2.14   REGULATORY COMPLIANCE IN IIoT RISK MITIGATION

Regulatory compliance plays a pivotal role in the risk mitigation strategies within the IIoT landscape. Companies adopting IIoT technologies must adhere to industry-specific regulations to ensure the secure and ethical deployment of connected devices. By monitoring and aligning IIoT practices with regulatory benchmarks, organizations can identify potential vulnerabilities and establish robust measures to

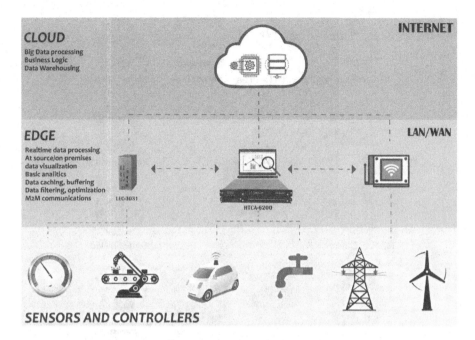

CLOUD
Big Data processing
Business Logic
Data Warehousing

INTERNET

EDGE
Realtime data processing
At source/on premises
data visualization
Basic analitics
Data caching, buffering
Data filtering, optimization
M2M communications   LEC-3031

LAN/WAN

HTCA-6200

SENSORS AND CONTROLLERS

FIGURE 3.9 Representation of edge and cloud computing [22].

address root causes. The digitization of power systems and the integration of IIoT sensors contribute to improved compliance and risk management in sectors such as power management and metering. This convergence of IIoT technologies and compliance measures enhances overall cybersecurity and operational resilience, ensuring that IIoT implementations meet the necessary standards and regulations.

3.2.15 RESILIENCE STRATEGIES FOR IIoT NETWORKS

Resilience in IIoT networks is crucial for maintaining operational continuity and mitigating potential risks. Various strategies focus on enhancing the robustness of IIoT systems. This includes implementing cyber resilience measures tailored for industrial networks, such as intrusion detection systems and secure communication protocols. Furthermore, building resilient industrial networks with IIoT involves addressing challenges through the deployment of IoT solutions, improving worker safety, increasing uptime, and enhancing operational efficiency. Learning from the lessons of the pandemic, organizations emphasize the ability of IIoT networks to withstand shocks, adapt to changing conditions, and recover swiftly from disruptions, emphasizing the significance of resilience in ensuring the long-term success of IIoT implementations. Figure 3.10 represents a resilient IIoT network.

3.2.16 DATA PRIVACY CONCERNS IN IIoT RISK MANAGEMENT

Data privacy concerns in IIoT risk management revolve around the collection, processing, and safeguarding of sensitive information within the IIoT ecosystem. The

FIGURE 3.10 Resilient IIoT networks [23].

sheer volume of data generated by IIoT devices poses a challenge, as it increases the risk of unauthorized access and potential breaches of personal information. The specific considerations required to protect individual's information in the IIoT environment emphasize the need for robust security measures, as IoT devices often collect and process user data without implementing adequate data protection measures. Addressing these concerns involves implementing stringent security protocols, ensuring compliance with regulations, and raising awareness about the importance of data privacy in IIoT deployments.

3.2.17 SUPPLY CHAIN VULNERABILITIES IN IIoT

Figure 3.11 represents a supply chain management framework in Industry. Supply chain vulnerabilities in IIoT refer to the risks associated with the interconnected network of devices and components involved in the manufacturing and distribution of IIoT systems.

These vulnerabilities can manifest at various stages, from the procurement of hardware components to the deployment of firmware and software. Issues such as counterfeit hardware, insecure firmware, and the potential for remote exploitation of IoT devices contribute to the complexity of IIoT supply chain security. The rapid expansion of the IIoT landscape intensifies these risks, emphasizing the need for robust measures to secure every link in the supply chain, from hardware manufacturing to software development and deployment.

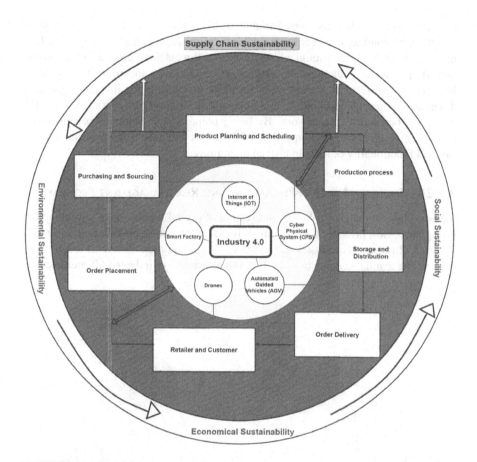

FIGURE 3.11 Supply chain management framework in industry [24].

3.2.18 HUMAN FACTORS IN IIoT RISK MITIGATION

Human factors play a crucial role in the effective mitigation of risks associated with IIoT systems. As IIoT increasingly integrates intelligent devices and connectivity into industrial processes, the human element becomes a pivotal aspect of risk management. This involves not only addressing technical vulnerabilities but also considering how human interactions, decision-making, and behaviors impact the overall security landscape. Factors such as employee training, awareness programs, and fostering a security-conscious culture contribute significantly to mitigating risks in IIoT implementations. Recognizing the human dimension in IIoT risk mitigation is essential for implementing comprehensive strategies that safeguard against potential threats and vulnerabilities.

3.2.19 INCIDENT RESPONSE PLANNING FOR IIoT SECURITY

Incident response planning is a critical component of securing IIoT environments. As IIoT systems become integral to industrial processes, the need for a well-defined incident response plan becomes paramount. Such a plan involves outlining steps to detect, respond to, and recover from security incidents that may impact the IIoT infrastructure. The plan should address potential threats and vulnerabilities specific to IIoT, providing guidelines for identifying, containing, and mitigating security breaches. Organizations should establish clear procedures for incident reporting, designate responsible individuals, and integrate incident response strategies into the broader cybersecurity framework. By being prepared for incidents, organizations can minimize the potential impact on IIoT devices and ensure the continued integrity and functionality of their industrial operations.

3.2.20 INDUSTRY 4.0 TRANSFORMATIONS AND RISK MANAGEMENT PROGRAMS

Industry 4.0, characterized by the integration of digital technologies into manufacturing processes, presents a paradigm shift that offers both opportunities and risks. Figure 3.12 represents Industry 4.0 concept using integration of several recent techniques. The seamless connectivity and data-driven nature of Industry 4.0 open avenues

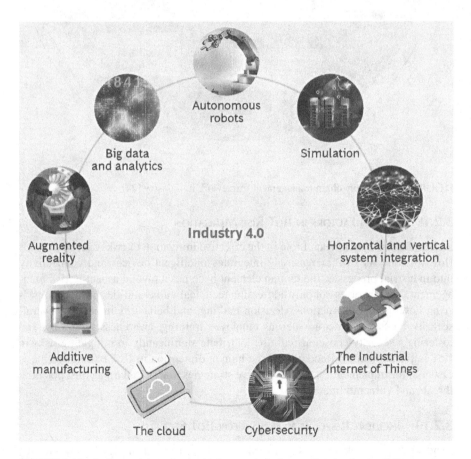

FIGURE 3.12 Industry 4.0 concept with integration of several technologies [25].

for enhanced efficiency and innovation. However, this transformation introduces new challenges, particularly in the realm of cybersecurity and data privacy. As organizations embark on Industry 4.0 initiatives, implementing robust risk management programs becomes imperative. These programs need to address cyber threats, data breaches, and the complexities of managing interconnected systems. Industry leaders recognize that reaping the rewards of Industry 4.0 technologies requires a strategic approach to risk management, safeguarding digital transformations against potential disruptions, and ensuring the resilience of smart manufacturing environments.

3.3   RECOMMENDED STEPS FOR RISK MANAGEMENT IN IIoT

Implementing robust risk management in the IIoT is essential for safeguarding critical infrastructure. Organizations should conduct a thorough Asset Inventory to catalog all IIoT devices and systems. Following this, regular risk assessments are crucial to identify vulnerabilities and potential threats. Continuous monitoring of IIoT networks is essential for real-time threat detection. It is imperative to implement data encryption for end-to-end protection of sensitive information and establish strict access control measures to prevent unauthorized entry. Vetting vendors for security standards and developing a comprehensive incident response plan are key components. Employee training on cybersecurity best practices enhances overall security awareness, while staying compliant with relevant regulations and standards is vital. Securing the IIoT supply chain and ensuring timely application of patch management further fortify defenses.

Physical security measures, including access restrictions, contribute to safeguarding IIoT infrastructure. Network segmentation limits lateral movement of threats, and staying informed through threat intelligence feeds adds an extra layer of defense. Lastly, regular security audits evaluate the effectiveness of implemented risk management measures. The important recommended steps for risk management in IIoT are listed below:

- Asset Inventory: Create a detailed inventory of all IIoT devices and systems to understand the attack surface.
- Risk Assessment: Conduct regular risk assessments to identify potential vulnerabilities and threats.
- Continuous Monitoring: Implement continuous monitoring of IIoT networks for real-time threat detection.
- Data Encryption: Ensure end-to-end encryption for data in transit and at rest to protect sensitive information.
- Access Control: Implement strict access controls and authentication mechanisms to prevent unauthorized access.
- Vendor Assessment: Assess and vet vendors supplying IIoT components for security standards.
- Incident Response Plan: Develop and regularly update an incident response plan to mitigate the impact of security incidents.
- Employee Training: Train employees on cybersecurity best practices to enhance overall security awareness.

- Regulatory Compliance: Stay compliant with relevant regulations and standards to avoid legal and regulatory issues.
- Supply Chain Security: Secure the IIoT supply chain to prevent tampering or compromise of components.
- Patch Management: Keep IIoT devices and systems up to date with the latest security patches.
- Physical Security: Implement physical security measures to protect IIoT infrastructure from unauthorized physical access.
- Network Segmentation: Segment networks to contain potential breaches and limit lateral movement of threats.
- Threat Intelligence: Stay informed about emerging threats and vulnerabilities through threat intelligence feeds.
- Security Audits: Conduct regular security audits and assessments to evaluate the effectiveness of risk management measures.

This multifaceted approach helps organizations navigate the evolving threat landscape and ensures the resilience of their IIoT ecosystems. These steps collectively contribute to building a robust risk management framework for IIoT, ensuring the resilience of industrial systems in the face of evolving cyber threats.

3.4   CONCLUSION

The landscape of risk management in the IIoT is a complex and dynamic arena. The review of existing literature reveals the multifaceted nature of cybersecurity challenges in IIoT ecosystems. Technologies and frameworks for cyber risk management have evolved to address the unique vulnerabilities associated with interconnected industrial devices. The four-layer IoT cyber risk management model, as discussed in the literature, offers a structured approach for organizations to navigate these challenges. The integration of cloud services, as exemplified by AWS, brings forth both opportunities and challenges in managing the vast amount of data generated by dispersed devices. Additionally, the paper highlights the significance of understanding environmental-based security risks and vulnerabilities in IIoT systems. Privacy concerns, security standards, and frameworks for IoT have been scrutinized to provide a comprehensive overview. As future developments continue to shape the IIoT landscape, it becomes imperative for organizations to remain vigilant, adopting proactive measures outlined in the literature. Continuous education, adherence to security standards, and a robust incident response plan are paramount for mitigating risks. This chapter serves as a valuable resource for researchers, practitioners, and decision-makers aiming to enhance the resilience of IIoT systems against evolving cybersecurity threats.

REFERENCES

1. Lee In, "Internet of things (IoT) cybersecurity: Literature review and IoT cyber risk management", Future Internet, 2020, 12, 157. https://doi.org/10.3390/fi12090157.
2. Zakaria, Huraizah, Abu Bakar, Nur Azaliah, Hafizah Hassan, Noor, Yaacob, Suraya, "IoT security risk management model for secured practice in healthcare environment",

Procedia Computer Science, 2019, 161, 1241–1248. https://doi.org/10.1016/j.procs. 2019.11.238.

3. Trinetra T-Sense, "Some Key Risks that Affect IioT Projects and Applications", Available online: https://www.trinetratsense.com/blog/key-risks-that-affect-iiot-projects-applications/ (accessed on 2 January 2024).

4. Adaros-Boye, C., Kearney, P., Josephs, M., Continuous Risk Management for Industrial IoT: A Methodological View. In: Kallel, S., Cuppens, F., Cuppens-Boulahia, N., Hadj Kacem, A. (eds) Risks and Security of Internet and Systems. CRiSIS 2019. Lecture Notes in Computer Science, 2020, vol 12026. Springer, Cham. https://doi.org/10.1007/ 978-3-030-41568-6_3.

5. Popescu, T.M., Popescu, A.M., Prostean, G. "IoT security risk management strategy reference model (IoTSRM2)", Future Internet, 2021, 13, 148. https://doi.org/10.3390/fi13060148.

6. Adaros Boye, C., Kearney, P., Josephs, M. (Cyber-Risks in the Industrial Internet of Things (IIoT): Towards a Method for Continuous Assessment. In: Chen, L., Manulis, M., Schneider, S. (eds) Information Security. ISC 2018. Lecture Notes in Computer Science, vol 11060, 2018. Springer, Cham. https://doi.org/10.1007/978-3-319-99136-8_27.

7. Ahmed, Shams Forruque, Alam, Md. Sakib Bin, Hoque, Mahfara, Lameesa, Aiman, Afrin, Shaila, Farah, Tasfia, Kabir, Maliha, Shafiullah, GM, Muyeen, S.M., "Industrial internet of things enabled technologies, challenges, and future directions", Computers and Electrical Engineering, 2023, 110, 108847. https://doi.org/10.1016/ j.compeleceng.2023.108847.

8. Thanaa Saad AlSalem, Mohammed Amin Almaiah, Abdalwali Lutfi, "Cybersecurity risk analysis in the IoT: A systematic review", Electronics, 2023, 12, 3958. https://doi. org/10.3390/electronics12183958.

9. IEC. IoT 2020: Smart and Secure IoT Platform. Available online: https://basecamp.iec.ch/ download/iec-white-paper-iot-2020-smart-and-secure-iot-platform/ (accessed on 2 January 2024).

10. World Economic Forum. State of the Connected World 2020 Edition INSIGHT REPORT. Available online: http://www3.weforum.org/docs/WEF_The_State_of_the_ Connected_World_2020.pdf (accessed on 2 January 2024).

11. McKinsey & Company. Cybersecurity in a Digital Era Your Guide to the Emerging Technologies Revolutionising Business Now. Available online: https://www.mckinsey. com/business-functions/risk/our-insights/cybersecurity-in-a-digital-era (accessed on 2 January 2024).

12. AT&T. AT&T Cybersecurity Insights: The CEO's Guide to Securing the Internet of Things. Available online: https://www.business.att.com/content/dam/attbusiness/insights/ migrated/exploringiotsecurity.pdf (accessed on 2 January 2024).

13. McKinsey & Company. How CEOs Can Tackle the Challenge of Cybersecurity in the Age of the Internet of Things. Available online: https://www.mckinsey.com/featured-insights/internet-of-things/our-insights/six-ways-ceos-can-promote-cybersecurity-in-the-iot-age (accessed on 2 January 2024).

14. Bain & Company. Cybersecurity Is the Key to Unlocking Demand in the Internet of Things. Available online: https://www.bain.com/insights/cybersecurity-is-the-key-to-unlocking-demand-in-the-internet-of-things/ (accessed on 2 January 2024).

15. World Economic Forum. Future Series: Cybersecurity, Emerging Technology and Systemic Risk Insight Report. Available online: http://www3.weforum.org/docs/WEF_ Future_Series_Cybersecurity_emerging_technology_and_systemic_risk_2020.pdf (accessed on 2 January 2024).

16. Ponemon Institute. A New Roadmap for Third Party IoT Risk Management the Critical Need to Elevate Accountability, Authority and Engagement. Available online: https:// sharedassessments.org/blog/a-new-roadmap-for-third-party-iot-risk-management/ (accessed on 2 January 2024).

17. National Academy of Engineering. NAE Grand Challenges For Engineering. Available online: http://www.engineeringchallenges.org/challenges/11574.aspx (accessed on 2 January 2024).
18. Sprintzeal, IoT Security Challenges and Best Practices-An Overview. Available online: https://www.sprintzeal.com/blog/iot-security-challenges (accessed on 2 January 2024).
19. Saleh, Z., Refai, H., Mashhour, A., "Proposed framework for security risk assessment," Journal of Information Security, 2011, 2 2, 85–90. https://doi.org/10.4236/jis.2011.22008.
20. Hazra, Abhishek, Adhikari, Mainak, Amgoth, Tarachand, Srirama, Satish Narayana,. "A comprehensive survey on interoperability for IIoT: Taxonomy, standards, and future directions", ACM Computing Surveys, 2021, 55, 1, 1–35. https://doi.org/10.1145/3485130.
21. Peter Reynolds, The Convergence of Artificial Intelligence and Industrial IoT, Available online: https://www.arcweb.com/industry-best-practices/convergence-artificial-intelligence-industrial-iot (accessed on 2 January 2024).
22. Kumar, Ravi, As an IoT Platform, What Should be the Right Balance of Data Computing Between the Edge and the Cloud? Available online: https://medium.com/world-of-iot/as-an-iot-platform-what-should-be-the-right-balance-of-data-computing-between-the-edge-and-the-ec6ea99344c8 (accessed on 2 January 2024).
23. Kumar, Gaurav, Building Resilient Industrial Networks with Industrial IoT. Softobotics, Available online: https://www.softobotics.com/blogs/building-resilient-industrial-networks-with-industrial-iot/ (accessed on 2 January 2024).
24. Zhang, G., Yang, Y., Yang, G. "Smart supply chain management in Industry 4.0: The review, research agenda and strategies in North America", Annals of Operations Research, 322, 2023, 1075–1117. https://doi.org/10.1007/s10479-022-04689-1.
25. Rüßmann, Michael, Lorenz, Markus, Gerbert, Philipp, Waldner, Manuela, Engel, Pascal, Harnisch, Michael, Justus, Jan. Industry 4.0: The Future of Productivity and Growth in Manufacturing Industries, Available online: https://www.bcg.com/publications/2015/engineered_products_project_business_industry_4_future_productivity_growth_manufacturing_industries/ (accessed on 2 January 2024).

4 Data Security for Industrial Internet of Things

An Essential Perspective for Industrial Engineering

Deepika Bhatia

4.1 OVERVIEW OF IIoT DATA SECURITY

Industrial Internet of Things (IIoT) is the use of smart devices such as sensors, meters, and smart faucets to automate industrial processes like healthcare, manufacturing, and agriculture. It leads to better accuracy, enhanced safety, less time, and money consumption. IIoT collects a huge amount of data and uses online storage for later analysis. Data can be uploaded to the cloud using servers or gateways, which leads to various security breaches. The intensive growth of the IIoT has opened a new era of connectivity and efficiency, but it has also exposed a myriad of challenges in data security. A lot of devices are connected using Internet of Things (IoT), including healthcare, aviation, and automation industries. IoT is a system of interrelated devices connected to the internet to transfer data to and from. Example: A smart home, where home general devices can be controlled to turn on the TV at a particular time, to open the door at a certain time of the day, etc. Cloud stores the data from these home devices and can be used for data analysis and knowledge extraction. As there is no human intervention, so, operational efficiency is increased due to lack of human error. There are various concerns for IoT security like cybersecurity features, activities, skills, capabilities, and IoT identity.

4.1.1 TRENDS AND CHALLENGES IN DATA SECURITY FOR IIoT DEVICES

4.1.1.1 Trends

In this topic, we delve into the emerging landscapes of data security for IIoT devices, exploring the latest trends and the unique hurdles that organizations need to navigate to safeguard their valuable information. As organizations embrace the potential of the IIoT, they must also address the key challenges of data security. The increasing complexity and interconnectedness of devices present vulnerabilities that can be exploited by cyber attackers. Robust encryption protocols should be implemented, conduct security audits regularly, and staying updated in regard to emerging threats

DOI: 10.1201/9781003466284-4

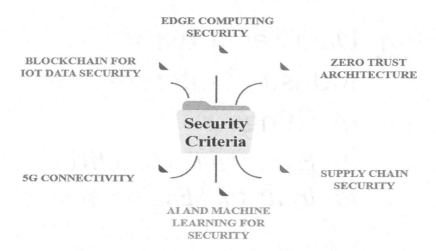

FIGURE 4.1 Trends in IIoT data security.

are crucial steps for safeguarding sensitive data in this rapidly evolving landscape. There are various trends as shown in Figure 4.1 above:

The IIoT presents various trends and challenges in terms of security. Trends are discussed in Table 4.1 below:

4.1.1.2 Challenges

IoT has revolutionized the idea that various devices communicate and share information, enabling a seamless interconnected environment. However, adoption of IoT technologies [14] has also given rise to significant challenges, particularly in the realm of data security are given below:

1. **Device Proliferation and Diversity:** The sheer number and diversity of IIoT devices which makes it challenging to implement consistent data security measures across the ecosystem. Managing security for a wide range of devices with varying capabilities and vulnerabilities is a complex task.
2. **Integrating Legacy System:** Many industrial systems have legacy components that may lack modern security features. Integrating these legacy systems with new IIoT technologies without compromising security is a significant challenge.
3. **Data Privacy Concerns:** IIoT devices often collect and process sensitive data. Ensuring compliance with data privacy standards, rules and regulations and protecting data from unauthorized access is a persistent challenge. Fog computing and cloud computing [15] techniques are combined by the researchers and they proposed a framework that can enhance the efficiency and security of data storage and retrieval in case of IIoT devices.
4. **Lack of Standardization:** The absence of standardized security protocols across the IIoT landscape can create interoperability issues and make it difficult to establish a unified security framework.
5. **Inadequate Authorization and Authentication:** Weak authentication mechanisms and insufficient authorization practices can lead to unauthorized access.

TABLE 4.1
Recent Trends in Security

Security Criteria	Trend
Edge Computing Security	Edge computing in IIoT is gaining popularity, enabling data processing closer to the source, enhancing efficiency, but necessitating robust security measures to protect sensitive data. Researchers proposed a new architecture for SDN and edge computing [1,2]. The architectures use lightweight load balancing algorithms but are vulnerable to DDoS attacks.
Zero Trust Architecture	Zero Trust [3] is gaining popularity for its continuous authentication and authorization model, ensuring security in dynamic IIoT environments regardless of the device's location or network. With new advancements in 5G-enabled IoT devices, there is a big issue of resource access and its grant. The researcher Zhang et al. [4,5] focused on new security approaches as compared to earlier traditional perimeter-based approaches to ensure data security from various types of attacks on resources by breaching the parameters. Researchers [6,7] proposed a zero trust architecture to never trust and verify everyone. A blockchain-based architecture [8] was proposed to ensure data security.
AI and Machine Learning for Security	AIML techniques are being utilized in IIoT environments for real-time anomaly detection and threat prediction, enhancing it to identify emerging data security threats.
Blockchain for IoT Data Security	Blockchain is being explored for improving data integrity and trust in IIoT by providing a decentralized, tamper-resistant ledger for secure transactions and data sharing among devices. Researchers [9] used BlackFog and BlackCloud platforms for AI-based attack detection in blockchain. Distributed ledgers [10] enable data security and are gaining popularity nowadays. Distributed ledgers are databases spread across multiple locations, requiring authentication from multiple instances to make any changes. This enhances security by creating shared accountability among network actors. Distributed ledger technologies (DLTs) are closely connected with IoT platforms and are used to efficiently manage data with a focus on security, privacy, and safety [11].
5G Connectivity	The implementation of 5G networks [4,5] will make connectivity for IIoT devices faster and more reliable, but it will also raise new security concerns, such as data protection and potential infrastructure vulnerabilities. Various types of threats [6] are possible while data communication via such networks, like IP Spoofing, Side Channel Attack, DDoS, vulnerability [12] of protocol and control nodes.
Supply Chain Security	The focus is on securing the entire IIoT device supply chain, including component integrity, vendor security practices, and securing update mechanisms. The researchers [13] comprehended new emerging technologies, like blockchain, machine learning, and physically unclonable functions, as solutions to different vulnerabilities present in the present infrastructure of the supply chain in IoT devices.

Strengthening identity management and access controls is critical for preventing unauthorized entry into IIoT systems.

6. **Architectural Challenges:** Nowadays, companies depend upon cloud databases as client-server architecture for data storage, processing, and transfer. It is prone to various issues like costing, computing resources, data

storage, device authentication, power consumption, delayed response, and data security. Edge and Distributed architectures [16] are gaining popularity nowadays and prove to be better in case of IoT security.

7. **Insufficient Update Mechanisms:** Updating firmware and software on IIoT devices is crucial for addressing vulnerabilities. However, many devices lack robust update mechanisms, leaving them exposed to known exploits.

Other challenges or issues are remote access, resource constraints, interoperability, real-time operations, physical safety, reliability, lack of awareness, and proper training. Addressing all these challenges and staying abreast of emerging trends is essential for organizations looking to secure their IIoT ecosystems effectively. Regularly updating security practices and adopting a proactive approach to risk management are key strategies in this dynamic and evolving landscape.

4.1.1.3   Approaches to IIoT Security

IIoT is different from IoT (used basically for consumers in commercial sector) as IIoT main purpose is in manufacturing and supply chain management systems generating voluminous data using cloud computing and machine learning, like smart factories, smart cities, smart agriculture, smart parking, and smart waste management system. IIoT uses more sensitive and precise controllers including location-aware technologies. The major issue in today's era is securing IoT and IIoT environments are crucial to protect critical infrastructure and sensitive data. The approaches mentioned below cover various aspects of IIoT security, and integrating them can create a comprehensive security framework. These approaches are given below:

1. **Robust Encryption and Authentication:**
   - Organizations should implement strong encryption algorithms for the protection of data during transmission.
   - Delegated authorization [14] and authentication can be considered in case of IoT device security.
   - Use different communication protocols (e.g., TLS/SSL), MQTT, HTTP, etc. which are secured enough to ensure the confidentiality and integrity of data.
   - Enforce robust authentication techniques, such as multi-factor authentication to control access to IoT devices and systems.
2. **Network Segmentation and Isolation:**
   - Divide the IIoT network into isolated segments so that potential impact of security breaches can be minimized.
   - Use firewalls and routers to control traffic between network segments.
   - Implement Virtual LANs to segregate different types of devices and limit lateral movement for attackers.
3. **Device Management:**
   - Employ a robust device management system to monitor, update, and patch IIoT devices regularly.
   - Ensure that devices have the latest firmware and security patches.
   - Implement secure boot process and device integrity verification.

- Unlike iOS, Android is open-sourced [14] leading to various vulnerabilities and attacks. Thus, its manufacturers should enrich IoT technology in such devices.

4. **Anomaly Detection and Intrusion Prevention Systems (IPS):**
   - Companies should deploy intrusion detection and prevention systems to monitor network traffic and control the suspicious activities coming inside and outside the IoT devices.
   - Use anomaly detection techniques to identify unusual behavior that may indicate a security threat in the device.
   - Implement real-time alerting and response mechanisms to address security incidents promptly.

5. **Regulatory Compliance:**
   - Stay compliant with relevant industry and regional regulations (e.g., GDPR, NIST, IEC 62443) to ensure a baseline level of security.
   - The researchers [14] focused on the use of lightweight cryptosystems and various security protocols as compared to earlier security protocols to reduce the computational overheads.
   - Regular audits can be done to assess compliance to identify and monitor any gaps in security practices.

6. **Employee Training:**
   - Provide regular training for employees to raise awareness about security best practices.
   - Educate staff on social engineering tactics and discuss among them the importance of adhering to security policies.
   - To develop a security-conscious culture within the organization.

7. **Incident Response Planning:**
   - Develop a systematic incident response plan that outlines steps to be taken in the event of a security incident.
   - Conduct regular drills and simulations to test the effectiveness of the incident response plan.
   - Establish communication protocols for notifying relevant stakeholders and authorities.

8. **Encourage ethical hacking:**
   - Manufacturers should encourage ethical hacking in their organizations. Rewards can be given and strict legislation should be implemented.

By integrating these approaches, organizations can create a robust IIoT security posture that addresses various aspects of cyber threats and vulnerabilities. Regular updates and continuous improvement are essential to adapt to evolving security challenges in the dynamic landscape of IIoT.

4.1.1.4   Security Algorithms for IIoT Data Security

Alternatively, to ensure the security of digital IoT systems, lightweight cryptographic techniques were used to minimize the potential threats, thereby increasing the safety of the system. Over the decades, many encryption algorithms, such as symmetric and asymmetric ones have been used enormously for securing data, information, and devices. A list of IIoT device security algorithms is given below in Table 4.2.

TABLE 4.2

Categories of Security Algorithms

Sr. No.	Algorithms Category	List of algorithms	Usage of Algorithms
1	Key Exchange Protocols	Diffie–Hellman Key Exchange (DHKE), Elliptic curve Diffie–Hellman Key Exchange.	Both are used for key exchange between networks in a secured manner.
2	Hash Functions	SHA-256 and MD5	Used for data security. Cryptographic hashing algorithms like SHA-256 are used to generate and verify checksums, ensuring the authenticity of firmware.
3	Digital Signatures	Elliptic Curve Digital Signature Algorithm (ECDSA), Rivest–Shamir–Adleman (RSA)	For securing digital signatures.
4	Secure Hash Algorithms for HMAC	Hash-Based Message Authentication Code (HMAC)	It combines a cryptographic hash function with a secret key for the verification of the integrity and authenticity of a message sent or received.
5	TLS/SSL Protocols	Transport Layer Security (TLS) and Secure Sockets Layer (SSL)	Use a combination of symmetric and asymmetric encryption algorithms for secure data transfer.
6	Public Key Infrastructure (PKI) Algorithms	X.509 Certificates, RSA, and Elliptic Curve Cryptography (ECC)	RSA and ECC are used for Key pair generation. Hardware-based cryptographic modules or dedicated co-processors may be employed for efficient PKI operations.
7	Random Number Generators	Secure Pseudo-Random Number Generators (PRNGs)	It is used for generating cryptographic keys and initialization vectors.
8	Blockchain Algorithms	Consensus Algorithms	Used for validation of transactions in blockchain, Proof of Work, Proof of Stake algorithms are used. Researchers [9] combined a lightweight blockchain with the Convivial Optimized Sprinter Neural Network-based AI mechanisms to improve the security of IIoT devices.
9	Symmetric (secret key encryption) and Asymmetric (public key encryption) Encryption Algorithms	Symmetric: Advanced Encryption Standard (AES), Data Encryption Standard (DES), Triple Data Encryption Standard (3DES), Asymmetric: RSA, DSA, and ECC	RSA is used for key exchange and digital signatures in securing communications. RSA is trusted for the data in transit over the internet, like SSL/TLS, S/MIME, SSH, cryptocurrencies, etc. ECC uses shorter key lengths as compared to RSA algorithm and therefore is better in case of IoT enabled devices as it used limited computing resources. DSA is used for digital signature encryption and used authentication mechanism.

(Continued)

TABLE 4.2 *(Continued)*
Categories of Security Algorithms

Sr. No.	Algorithms Category	List of algorithms	Usage of Algorithms
10	Lightweight Cryptographic Algorithms [17]	PRESENT, PRINCE, CLEFIA, PICCOLO, and LBLOCK [18–22] are a few of the lightweight cryptographic algorithms currently used in IoT applications in a testing environment.	These are lightweight cryptographic algorithms. Most of them use Feistel/SPN architecture.

Various algorithms have been implemented in various IoT applications. However, choosing a suitable algorithm has always been a challenge. Additionally, regular updates and patches are crucial to address emerging threats and vulnerabilities in the rapidly evolving security landscape.

4.1.1.5 Computational Intelligence and Machine Learning Algorithms for IIoT Data Security

The traditional defense systems in IIoT devices arc incapable of detecting new versions of cyber-attacks. The CI techniques provide various benefits for the IoT that connect vast number of smart devices with each other. There are various computational intelligence algorithms [23] for IoT data security, such as Fuzzy logics, immune systems, expert systems, swarm intelligence, evolutionary computations, and artificial intelligence networks [10].

Nowadays, AI techniques such as ANN, machine learning, deep learning (DL), Fuzzy logic, particle swarm optimization (PSO) techniques can be combined with Robotics [24] and IIoT also, which helps the devices to perform activities using intelligence via AI and do tasks in an autonomous way. Various DL-based techniques are available in the market for network intrusion detection. DL technique extracts hidden patterns from the network traffic. It extracts suitable threat patterns automatically, assisting in the understanding and detection of cyber threats, as well as providing appropriate intelligence to classify the type of attack. Also, PSO can be combined with deep RNN to provide security in IIoT devices.

Researchers [23] focused on the applications of computational intelligence techniques in cybersecurity for IoT devices, CI-based cyber defending technologies, CI-based tools, CI-enabled data mining in cybersecurity, intrusion detection techniques, and various challenges and issues in data security in IoT. Fog computing and cloud computing [15] techniques are combined by the researchers and a framework was proposed that can greatly improve the efficiency and security of data storage and retrieval in IIoT.

There are various machine learning algorithms that can be used for device authentication. For securing communication signals [16], PCA, ANN, and random forest algorithms are used by various researchers [25], ensemble learning is used for

TABLE 4.3

Machine Learning Algorithms for IoT Security

Sr. No.	ML Classification	ML Technique
1	Supervised Learning [28]	Decision Trees, Naïve Bayes, SVM, and KNN are used to detect DoS/DDoS attacks.
2	Unsupervised Learning [29]	K-means, PCA are prone to DoS/DDoS attacks.
3	Deep Learning [30]	CNN and RNN are used for device intrusion and malware detection. Researchers [31] discussed intrusion detection systems, including RNN, SVM, K-means, SOM, LR, and RF mechanisms.
4	Reinforcement Learning [32]	Uses sequential decision.
5	Markov Process	Uses sequential decision.
6	Ensemble Learning [33]	Bragging and boosting methods.

network traffic that uses white lists [26], decision tree, logistic regression, and Naïve Bayes classifiers are used for securing operation behavior and device information using dynamic fingerprint technology in IoT devices [27]. The list of various decision algorithms is given below in Table 4.3.

Deep Neural Network [34] based models are like black boxes; they learn from changing parameters and draw conclusions. It gives low transparency and poor interpretation of results.

4.2 CONCLUSION

IIoT data security is crucial for organizations navigating interconnected industrial devices. It involves encryption, authentication, network segmentation, device management, anomaly detection, regulatory compliance, employee training, and incident response planning. Continuous adaptation to security challenges and best practices ensures confidentiality, integrity, and availability of critical data. As the IoT landscape continues to evolve, an adaptive and proactive approach to data security is essential to mitigate risks and ensure the long-term success of IoT deployments.

REFERENCES

1. K. Kalkan and S. Zeadally, "Securing Internet of Things with Software Defined Networking," *IEEE Communications Magazine*, vol. 56, no. 9, pp. 186–192, 2018, doi: 10.1109/MCOM.2017.1700714.
2. D. Yin, L. Zhang and K. Yang, "A DDoS Attack Detection and Mitigation with Software-Defined Internet of Things Framework," *IEEE Access*, vol. 6, pp. 24694–24705, 2018, doi: 10.1109/ACCESS.2018.2831284.
3. F. A. Qazi, "Study of Zero Trust Architecture for Applications and Network Security," in *2022 IEEE 19th International Conference on Smart Communities: Improving Quality of Life Using ICT, IoT and AI (HONET)*, Marietta, GA, USA, 2022, pp. 111–116, doi: 10.1109/HONET56683.2022.10019186.

4. J. Zhang, Y. Wang, S. Li and S. Shi, "An Architecture for IoT-Enabled Smart Transportation Security System: A Geospatial Approach," in IEEE Internet of Things Journal, vol. 8, no. 8, pp. 6205–6213, 2021, doi: 10.1109/JIOT.2020.3041386.
5. D. Zhang, Z. Zhou, S. Mumtaz, J. Rodriguez and T. Sato, "One Integrated Energy Efficiency Proposal for 5G IoT Communications," *IEEE Internet of Things Journal*, vol. 3, no. 6, p. 1346, 2016.
6. S. Bhattacharjya, *"A Novel Zerotrust Framework to Secure IoT Communications,"* Ph.D. thesis, University of Kansas, 2020.
7. H. Lin, J. Hu, W. Xiaoding, M. F. Alhamid and M Jalil Piran, "Toward Secure Data Fusion in Industrial IoT Using Transfer Learning," *IEEE Transactions on Industrial Informatics*, vol. 17, no. 10, pp. 1–1, 2020, doi: 10.1109/TII.2020.3038780.
8. S. Li, "Editorial: Zero Trust based Internet of Things" "EAI Endorsed Transactions on Internet of Things" vol. 5, no. 20, 2020.
9. Selvarajan et al., "An Artificial Intelligence Lightweight Blockchain Security Model for Security and Privacy in IioT Systems" *Journal of Cloud Computing*, 2023, doi: 10.1186/s13677-023-00412-y
10. S. Li, I. Muddesar and S. Neetesh, "Future Industry Internet of Things with Zero-trust Security" Information Systems Frontiers, Springer 2021, pp. 1–14, doi: 10.1007/s10796-021-10199-5.
11. A. Papageorgiou, T. Krousarlis, K. Loupos and A. Mygiakis, "DPKI: A Blockchain-Based Decentralized Public Key Infrastructure System," in *Global IoT Summit 2020, 3rdWorkshop on Internet of Things Security and Privacy (WISP) (Dublin)*, 2020, doi: 10.1109/GIOTS49054.2020.9119673.
12. J. Zhang, H. Chen, L. Gong, J. Cao and Z. Gu, "The Current Research of IoT Security," *2019 IEEE Fourth International Conference on Data Science in Cyberspace (DSC)*, Hangzhou, China, 2019, pp. 346–353, doi: 10.1109/DSC.2019.00059.
13. H. Vikas et al., "A Survey on Supply Chain Security: Application Areas, Security Threats, and Solution Architectures" *IEEE Internet of Things Journal*, vol. 8, no. 8, pp. 6222–6246, 2021, doi: 10.1109/JIOT.2020.3025775.
14. Z. -K. Zhang, M. C. Y. Cho, C. -W. Wang, C. -W. Hsu, C. -K. Chen and S. Shieh, "IoT Security: Ongoing Challenges and Research Opportunities," in *2014 IEEE 7th International Conference on Service-Oriented Computing and Applications*, Matsue, Japan, 2014, pp. 230–234, doi: 10.1109/SOCA.2014.58.
15. J.-S. Fu, Y. Liu, H. -C. Chao, B. K. Bhargava and Z. -J. Zhang, "Secure Data Storage and Searching for Industrial IoT by Integrating Fog Computing and Cloud Computing," *IEEE Transactions on Industrial Informatics*, vol. 14, no. 10, pp. 4519–4528, 2018, doi: 10.1109/TII.2018.2793350.
16. H. Wui et al., "Research on Artificial Intelligence Enhancing Internet of Things Security: A Survey" *IEEE Access*, vol. 8, pp. 153826–153848, 2020, Electronic ISSN: 2169-3536, doi: 10.1109/ACCESS.2020.3018170.
17. A. Bora et al., "Energy Consumption Analysis of Lightweight Cryptographic Algorithms That Can Be Used in the Security of Internet of Things Applications" *Hindawi Security and Communication Networks*, vol. 2020, pp. 1–15, 2020. doi: 10.1155/2020/8837671.
18. A. Bogdanov, L. R. Knudsen and G. Leander et al., "PRESENT: An ultra-lightweight block cipher," in Proceedings of the International Workshop on Cryptographic Hardware and Embedded Systems, September 2007.
19. T. Shirai, K. Shibutani, T. Akishita, S. Moriai and T. Iwata, "e 128-bit blockcipher CLEFIA," in Proceedings of the International Workshop on Fast Software Encryption, March 2007.
20. K. Shibutani, T. Isobe and H. Hiwatari et al., "Piccolo: An ultralightweight blockcipher," in Proceedings of the International Workshop on Cryptographic Hardware and Embedded Systems, September 2011.

21. J. Borghoff, A. Canteaut and T. G¨uneysu et al., "PRINCE–A Lowlatency Block Cipher for Pervasive Computing Applications," in *Proceedings of the International Conference on the 0eory and Application of Cryptology and Information Security*, December 2012. 14 Security and Communication Networks

22. W. Wu and L. Zhang, "LBlock: A lightweight block cipher," in Proceedings of the International Conference on Applied Cryptography and Network Security, June 2011.

23. Z. Shanshan, "Computational Intelligence Enabled Cybersecurity for the Internet of Things," *IEEE Transactions on Emerging Topics in Computational Intelligence*, vol. 4, no. 5, pp. 666–674, October 2020, Electronic ISSN: 2471-285X, doi: 10.1109/TETCI. 2019.2941757.

24. V. Ovidiu et al., "Internet of Robotic Things Intelligent Connectivity and Platforms", *Frontiers in Robotics and AI*, vol. 7, pp. 1–33, September 2020, doi: 10.3389/frobt.2020. 00104.

25. Y. Lin, X. Zhu, Z. Zheng, Z. Dou and R. Zhou, "The Individual Identification Method of Wireless Device Based on Dimensionality Reduction and Machine Learning," *Journal of Supercomputing*, vol. 75, no. 6, pp. 3010–3027, 2019.

26. Y. Meidan, M. Bohadana, A. Shabtai, M. Ochoa, N. O. Tippenhauer, J. D. Guarnizo and Y. Elovici, "Detection of unauthorized IoT devices using machine learning techniques," 2017, arXiv:1709.04647.

27. Z. Li, W. Zuoyue, W. Chundong, M. A. Yunfei and X. Chaocan, "Design and Implementation of Intelligent Identification System for IoT Terminals," *Journal of Chongqing University Posts Telecommunications*, vol. 31, no. 4, pp. 443–450, 2019.

28. R. Caruana and A. Niculescu-Mizil, "An Empirical Comparison of Supervised Learning Algorithms," in Proceedings of the 23rd International Conference on Machine Learning, 2006, pp. 161–168, doi: 10.1145/1143844.1143865.

29. B. C. Love, "Comparing Supervised and Unsupervised Category Learning," in *Psychonomic Bulletin & Review*, vol. 9, no. 4, pp. 829–835, 2002, doi: 10.3758/BF03196342.

30. M. Mohammadi, A. Al-Fuqaha, S. Sorour and M. Guizani, "Deep Learning for IoT Big Data and Streaming Analytics: A Survey," *IEEE Communication Surveys Tuts*, vol. 20, no. 4, pp. 2923–2960, 1st Quart., 2018, doi: 10.1109/COMST.2018.2844341.

31. H Bangui and B Buhnova, "Recent Advances in Machine-Learning Driven Intrusion Detection in Transportation: Survey," Procedia Comput Sci, vol. 184, pp. 877–886, 2021.

32. K. Gai and M. Qiu, "Optimal Resource Allocation Using Reinforcement Learning for IoT Content-Centric Services," *Applied Soft Computing*, vol. 70, pp. 12–21, 2018, doi: 10.1016/j.asoc.2018.03.056.

33. R. Polikar, "Ensemble Learning," *Ensemble Mach. Learn.*, vol. 4, pp. 1–34, 2012, doi: 10.1007/978-1-4419-9326-7_1.

34. D. Castelvecchi, "Can We Open the Black Box of AI?" Nature, vol. 538, no. 7623, pp. 20–23, 2016, doi: 10.1038/538020A.

5 Navigating Compliance for the Industrial IoT Landscape
Frameworks, Security, Standards, and Key Ethical Considerations

Ashima Shahi, Shaloo Bansal, Sunil Kumar Chawla, and Ahmed A. Elnger

5.1 INTRODUCTION

The Industrial Internet of Things (IIoT) stands as a transformative force in modern industry, extending the principles of the Internet of Things (IoT) to revolutionize industrial processes. Its foundation lies in fostering seamless communication between machines, harnessing the potential of big data analytics, and integrating machine learning algorithms [1]. Through these capabilities, the IIoT enables industries to achieve unprecedented levels of efficiency and reliability across various sectors, from robotics to healthcare to manufacturing [2]. This convergence of technology transcends traditional boundaries, merging information technology (IT) with operational technology (OT) to create a cohesive ecosystem that drives automation, optimization, and enhanced visibility throughout the supply chain and logistics networks [3].

Embedded within the broader context of the fourth industrial revolution, the IIoT represents a cornerstone of Industry 4.0's vision for interconnected cyber-physical systems [4]. Real-time data insights gleaned from sensors and information sources empower industrial machinery to make autonomous decisions, paving the way for levels of automation and efficiency previously unattainable [5]. Furthermore, the IIoT facilitates the evolution of smart cities and factories, where interconnected devices and intelligent systems orchestrate a symphony of operations [6]. By seamlessly exchanging data and insights, the IIoT empowers industries to uncover growth opportunities, optimize asset utilization, and make informed, data-driven decisions that propel them into the future.

This chapter comprehensively explores the multifaceted landscape of compliance in IIoT deployments. It begins by elucidating the organizational imperatives driving

IIoT adoption and the challenges associated with ensuring compliance in complex industrial environments. Subsequent sections delve into compliance frameworks tailored for IIoT, strategies for data security and privacy compliance, the role of standards and protocols, ethical considerations in artificial intelligence (AI) implementation, and the legal and regulatory implications of edge computing in IIoT environments. By addressing these diverse aspects, the paper provides a holistic understanding of compliance considerations in IIoT initiatives, offering valuable insights and best practices to navigate regulatory complexities and achieve compliance excellence in IIoT deployments.

5.1.1 ORGANIZATIONAL NEED FOR IIoT

IIoT shown in Figure 5.1 offers organizations a compelling proposition for advancing their operations in today's rapidly evolving industrial landscape. By integrating IIoT solutions into their infrastructure, businesses gain access to a wealth of actionable data that can drive transformative changes. This data isn't just about monitoring; it is about leveraging insights to enhance various aspects of operations. From improving worker safety through real-time monitoring to optimizing production uptime via predictive maintenance, IIoT enables organizations to stay ahead of the curve.

Furthermore, IIoT holds the key to unlocking operational efficiencies and maintaining regulatory compliance while preserving product quality. By streamlining processes, automating data collection, and facilitating real-time responsiveness, IIoT empowers organizations to navigate challenges with agility and precision. Embracing IIoT isn't merely an option for modern businesses—it is a strategic imperative for those seeking to thrive amidst technological disruption and drive sustainable growth in an increasingly competitive market landscape.

5.1.2 CHALLENGES IN COMPLIANCE OF IIoT

Compliance in the context of IIoT refers to the adherence of IIoT systems, devices, and processes to relevant laws, regulations, industry standards, and internal policies.

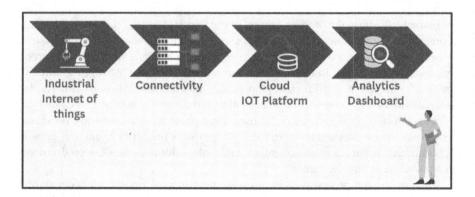

FIGURE 5.1 IIoT architecture.

Achieving compliance in IIoT is crucial for ensuring the security, privacy, reliability, and integrity of data and operations within industrial environments. As IIoT systems become increasingly integral to critical infrastructure and industrial processes, compliance requirements play a critical role in protection against potential risks and vulnerabilities.

The requirements for compliance in IIoT encompass various aspects, including but not limited to:

a. Data Security and Privacy: IIoT devices collect and transmit huge amounts of data, containing sensitive information related to operations, machinery, and personnel. Compliance mandates stringent solutions to secure this data from illegal access, threats, breaches, and misuse. Requirements for data security and privacy may comprise encryption protocols, access control methods, data anonymization techniques, and adherence to privacy protocols such as the General Data Protection Regulation (GDPR) or the Health Insurance Portability and Accountability Act (HIPAA).

b. Regulatory Standards and Guidelines: Compliance in IIoT necessitates adherence to relevant regulatory frameworks and industry guidelines specific to the sector or geographic region. These standards may cover aspects such as safety, environmental regulations, product certifications, quality management systems (e.g., ISO 9001), and cybersecurity protocols (e.g., National Institute of Standards and Technology (NIST) cybersecurity framework).

c. Interoperability and Compatibility: IIoT devices often operate within complex ecosystems comprising diverse hardware, software, and communication protocols. Compliance requires ensuring interoperability and compatibility among these components to facilitate seamless integration, data exchange, and collaboration across the IIoT ecosystem.

d. Cybersecurity and Resilience: IIoT systems are susceptible to cybersecurity threats such as viruses, worms, ransomware, denial-of-service attacks, and insider threats. Compliance mandates robust cybersecurity measures, including intrusion detection system (IDS), network segmentation, regular security audits, incident response plans, and employee training programs to mitigate risks and ensure operational resilience.

e. Data Governance and Lifecycle Management: Compliance entails establishing comprehensive data governance policies and practices to govern the collection, storage, processing, and disposal of data throughout its lifecycle. This includes defining data ownership, retention periods, data access controls, audit trails, and mechanisms for data sanitization or destruction when no longer needed.

f. Supply Chain Integrity: IIoT deployments often involve multiple vendors and third-party suppliers, raising concerns about supply chain integrity and cybersecurity risks. Compliance mandates rigorous supplier assessments, contractual agreements, and due diligence processes to ensure the integrity and security of components, software, and services sourced from external entities.

g. Ethical and Legal Considerations: Compliance in IIoT extends beyond technical requirements to encompass ethical and legal considerations, including issues related to transparency, accountability, fairness, and responsible use of technology. Compliance efforts should align with ethical frameworks and principles to mitigate potential societal impacts and ensure the ethical stewardship of IIoT systems and data.

Compliance in the IIoT encompasses a multifaceted approach that addresses data security, regulatory standards, interoperability, cybersecurity, data governance, supply chain integrity, and ethical considerations. By meeting these requirements, organizations can build trust, mitigate risks, and unlock the full potential of IIoT technologies to drive innovation and growth in industrial environments.

5.2 COMPLIANCE FRAMEWORKS FOR IIoT DEPLOYMENTS

Compliance frameworks for IIoT deployments provide structured guidelines and best practices to ensure that IIoT systems, devices, and processes adhere to relevant regulatory requirements, industry standards, and organizational policies. These frameworks are essential for managing the complexity and risks associated with IIoT deployments and helping organizations navigate compliance challenges effectively. Here are details of some prominent compliance frameworks tailored for IIoT deployments:

a. NIST Cybersecurity Framework: This framework is developed by the NIST, as shown in Figure 5.2; it provides a comprehensive approach for managing cybersecurity risks [7]. It consists of a set of strategies, guidelines, standards, and best practices organized into five core functions namely identify, protect, detect, respond, and recover. This framework aids establishments evaluate and improve their cybersecurity posture, including IIoT systems, by identifying and mitigating cyber threats and vulnerabilities.

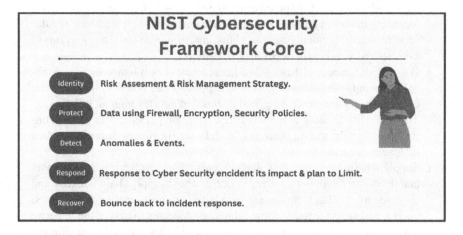

FIGURE 5.2   NIST cybersecurity framework.

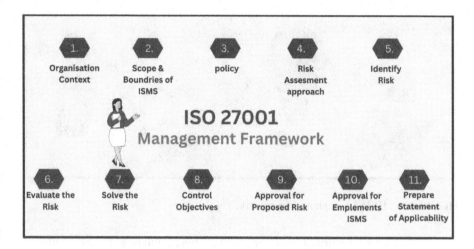

FIGURE 5.3 ISO/IEC 27001 management framework.

b. ISO/IEC 27001: Information Security Management System (ISMS): ISO/IEC 27001, shown in Figure 5.3 is an international standard that specifies requirements to establish, implement, maintain, and continuously improve an ISMS. It offers an organized way to manage crucial and sensitive information, including data collected and processed by IIoT devices and sensors. Compliance with ISO/IEC 27001 exhibits an organization's commitment to protect the CIA (confidentiality, integrity, and availability) triad of information assets, including those involved in IIoT deployments [8].

c. IEC 62443: Industrial Automation and Control Systems Security: Developed by the International Electrotechnical Commission (IEC), IEC 62443 is a series of standards and technical reports focused on cybersecurity for industrial automation and control systems (IACS), which include IIoT devices [9]. The standards provide guidelines for implementing cybersecurity measures tailored to the unique requirements of industrial environments, including risk assessment, network segmentation, access control, and incident response.

d. GDPR: GDPR, provided by the European Union (EU), is an all-inclusive data protection regulatory framework that applies to organizations processing personal data of individuals, including IIoT data [10]. This compliance requires organizations to ensure procedures to safeguard the legitimate, legal, and transparent processing of personal data, including data minimization, consent management, data subject rights, and data breach notification.

e. HIPAA: As shown in Figure 5.4, it is a U.S. federal law that sets principles for the security of complex personal health information, including healthcare-related IIoT data. Compliance with HIPAA requires organizations to implement organizational, physical, and technical precautionary rules to protect the CIA triad of protected health information (PHI) and other healthcare data [11].

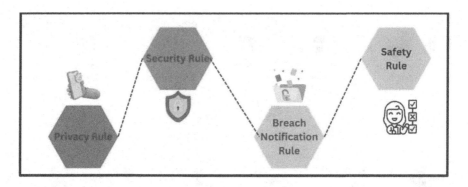

FIGURE 5.4   HIPAA compliance framework.

f. Industrial IoT Security Framework (IISF): This compliance framework and protocol was developed by the Industrial Internet Consortium (IIC) and is a detailed security framework specifically designed for IIoT deployments. It provides guidance and best practices for securing IIoT systems throughout the lifecycle, including device identification and authentication, secure communication, data protection, and security monitoring [12–14].

These compliance frameworks serve as valuable resources for organizations embarking on IIoT deployments, helping them establish robust security and privacy controls, mitigate risks, and demonstrate compliance with regulatory requirements and industry standards. By adopting and implementing these frameworks, organizations can build trust, increase resilience, and expose the full potential of IIoT technologies in commercial and industrial environments. Further, a tabular representation is given below for use case specific to the healthcare industry for each compliance framework (Table 5.1).

These challenges highlight the complexities and considerations organizations face when implementing compliance measures within the healthcare industry's IIoT deployments, underscoring the importance of robust security controls, risk management practices, and regulatory adherence.

5.3   DATA SECURITY AND PRIVACY COMPLIANCE IN IIoT

Strong data security is essential for the IIoT, as businesses depend more and more on networked devices and sensors to increase productivity. We will examine the best practices and regulatory requirements for handling data security in the IIoT environment. This section will cover the essential problem-solving techniques and factors for protecting your important data, from adhering to industry-specific requirements to putting strict access restrictions in place. We'll consider how crucial data transfer protocols, authentication, and encryption are to stopping illegal access and guaranteeing the privacy and accuracy of your personal data. Businesses can reduce the risk of data breaches by implementing best practices and comprehending the regulatory environment around data security [15,16].

TABLE 5.1

Compliance Framework for Healthcare Industry as a Use Case

Compliance Framework	Use Case	Benefits	Challenges
NIST Cybersecurity Framework	Secure remote access for telemedicine services	- Ensures secure and encrypted access to patient data for remote consultations	- Ensuring secure and reliable internet connectivity for remote consultations
	Cyber threat detection for medical devices	- Early detection of cybersecurity threats to prevent disruptions in healthcare	- Integration of cybersecurity measures with legacy medical device infrastructure
ISO / IEC 27001	Protection of electronic health records (EHR)	- Ensures confidentiality, integrity, and availability (CIA) of sensitive patient data	- Establishing and maintaining comprehensive access controls and user permissions for EHR systems
	Access control for healthcare personnel	- Restricts access to patient information to authorized healthcare professionals	- Managing access rights for temporary or rotating healthcare staff
IEC 62443	Securing medical IoT devices and wearables	- Enhances security measures to protect patient data and device functionality	- Ensuring compatibility and interoperability of security solutions with diverse medical IoT devices
	Network segmentation for hospital networks	- Isolates critical medical devices and systems from unauthorized access	- Balancing network segmentation with the need for seamless communication between healthcare systems
GDPR	Compliance with patient consent and data rights	- Ensures lawful and transparent processing of patient data	- Obtaining explicit consent from patients for data processing and sharing
	Data breach notification and response	- Timely notification and mitigation of data breaches to protect patient privacy	- Establishing efficient incident response procedures and communication channels in the event of a breach
HIPAA	Encryption of protected health information (PHI)	- Safeguards sensitive patient data from unauthorized access and disclosure	- Implementing encryption without compromising the performance of healthcare systems
	Audit trails for PHI access and disclosures	- Provides traceability and accountability for all accesses to patient records	- Ensuring the completeness and accuracy of audit logs for regulatory compliance

5.3.1 MANDATES GOVERNING DATA SECURITY

Understanding different legal criteria is necessary to ensure data security in the IIoT environment. Businesses need to efficiently traverse a complex web of regulations in order to protect sensitive data. GDPR is one such law that pertains to businesses that function within the EU or handle the personal data of people of the EU. Under the GDPR, businesses must put in place the proper organizational and technical safeguards to secure private data and notify individuals of any breaches as soon as they occur.

Aside from the GDPR, many sectors are subject to particular data security laws, e.g., in the United States, HIPAA governs the healthcare sector. Organizations must adhere to HIPAA's requirements for patient data protection, which include using audit logs, access controls, and encryption. Furthermore, businesses handling credit card data are subject to the Payment Card Industry Data Security Standard (PCI DSS). Implementing strong security measures, such as network segmentation, encryption, and secure coding techniques, is necessary to comply with PCI DSS [17].

Businesses must consider local data protection laws in the nations in which they conduct business in addition to these restrictions. The requirements of these regulations may differ, thus it is critical to stay current on the legislation to guarantee compliance.

5.3.2   BEST PRACTICES FOR ENHANCING SECURITY OF IIOT DEVICES AND NETWORKS

Ensuring the security of IIoT devices and networks is paramount for safeguarding sensitive data and thwarting unauthorized access [18,19]. Here are some recommended best practices:

a. Enforce robust access controls:
   Restrict access to critical systems and data to authorized personnel only. Utilize strong passwords, multi-factor authentication, and role-based access controls to bolster security.
b. Maintain regular device updates and patches:
   Stay vigilant against vulnerabilities by consistently updating devices with the latest firmware and security patches. This supports mitigating potential risks associated with unpatched vulnerabilities.
c. Implement network segmentation:
   Isolate IoT devices from the primary corporate network to contain potential breaches. Segmenting networks ensure that compromised devices have limited impact, restricting attackers' access.
d. Perform routine vulnerability assessments and penetration tests:
   Assess the security posture of IoT devices and networks through regular vulnerability assessments and penetration tests. Proactively identifying and addressing vulnerabilities can preemptively thwart potential attacks.
e. Employ network traffic monitoring and analysis:
   Deploy network monitoring tools to detect suspicious activity or anomalous behavior. Analyzing network traffic patterns enables the timely identification of security incidents, facilitating swift responses.

By following these best practices, organizations can bolster the security of their IIoT sensors, devices, and networks, removing the risk of data breaches and enhancing overall cybersecurity posture.

5.3.3   SECURING DATA THROUGH ENCRYPTION AND AUTHENTICATION

Encryption and authentication are two essential components of data security in the IIoT. Even in the event that it is intercepted or accessed, encryption makes sure that private information is unreadable by unauthorized parties. Conversely, authentication confirms the legitimacy of the devices and users, making system access possible. Protecting data while it is in transit and at rest can be achieved by putting strong encryption algorithms, such as the advanced encryption standard (AES), into practice. Many data security standards advocate AES, which is commonly considered to be quite secure [20,21].

Strong authentication procedures must be put in place in addition to encryption to guarantee that only authorized people and devices may access the system. Techniques like biometric authentication, secure key exchange protocols, and digital certificates can be used to accomplish this. Organizations should also think about deploying secure protocols for data transport, including secure shell (SSH) or transport layer security (TLS). By offering secure communication channels, these protocols guarantee the integrity and confidentiality of data while it is being transmitted [22].

Organizations may greatly improve the security of their IIoT systems and shield critical data from unwanted access by implementing encryption and authentication solutions.

5.3.4   ENHANCING DATA STORAGE AND TRANSMISSION SECURITY

In the IIoT landscape, safeguarding data storage and transmission is imperative to thwart unauthorized access and uphold data integrity [23]. Here are key best practices to consider:

a. Encrypt Data at Rest: Implement robust data encryption techniques to secure data stored in repositories. Utilize strong encryption algorithms and effective key management practices to ensure data remains inaccessible even if the storage medium is compromised.

b. Secure Data Transmission: Employ secure protocols like TLS or SSH for transmitting data between devices and systems. These protocols encrypt data during transit, preventing interception or tampering by unauthorized entities.

c. Opt for Secure Cloud Storage: Choose reliable cloud service providers (CSPs) offering robust security measures for IIoT data storage. Ensure encryption is applied for both data in transit and at rest, along with the stringent access controls and regular security audits.

d. Establish Regular Data Backup: Implement a comprehensive data backup strategy to facilitate swift data recovery in the event of a breach or system failure. Regularly test the backup and restoration processes to ensure their effectiveness.

e. Enforce Data Retention Policies: Define data preservation policies aligned with legal requirements and business objectives. Frequently review and remove outdated or needless data to minimize exposure risks.

By adhering to these practices, organizations can significantly enhance the security of their IIoT data, mitigating the likelihood of data breaches and unauthorized access.

5.3.5 MANAGING THREATS AND RESPONDING TO INCIDENTS

In the context of the IIoT, threat detection and incident response are essential to maintaining data security. To reduce the effect on their systems and data, organizations must be proactive in identifying and responding to possible security issues. Potential assaults and unusual activity can be found by putting strong threat detection measures in place, such as intrusion prevention systems (IPS) and IDS. These programs keep an eye on network activity, look for trends, and sound an alarm when something seems off [24].

Organizations should also create an incident response strategy (IRS) that describes what should be done in the case of a security occurrence. Procedures for containment, elimination, and recovery should be incorporated into this plan. To make sure the IRS is successful, test and update it frequently. Organizations should also think about cooperating on incident response initiatives and exchanging threat intelligence with colleagues in the industry. By working together, we can more successfully anticipate potential dangers and reduce risks. Organizations may reduce the effect of security incidents on their IIoT systems and data by investing in threat detection mechanisms and creating a strong incident response plan.

5.3.6 SELECTING OPTIMAL DATA SECURITY SOLUTIONS FOR IIoT

Choosing the right data security solutions for IIoT is crucial. Here's what to consider:

a. Scalability: Ensure the solutions can handle the increasing number of IoT devices and data volumes as your IIoT network grows. Look for solutions that can efficiently manage large-scale deployments without compromising performance or security.
b. Compatibility: Check if the data security solutions seamlessly integrate with your existing IoT devices, platforms, and communication protocols. Compatibility ensures smooth implementation and operation, minimizing disruptions to your current infrastructure.
c. Encryption and Authentication: Prioritize data security solutions that offer robust encryption algorithms and authentication mechanisms. Encryption protects data both at rest and in transit, while strong authentication ensures that only authorized users and devices can access sensitive information.
d. Monitoring: Choose solutions equipped with real-time monitoring and analytics capabilities. These features enable proactive detection of security threats and anomalous activities within your IIoT environment, allowing for swift response and mitigation.
e. Vendor Reputation: Select data security solutions from reputable vendors with a proven track record in the industry. Consider factors such as vendor reliability, product quality, and customer support services. Trusted vendors are more likely to provide regular updates, patches, and technical assistance to address security vulnerabilities and support your IIoT deployment effectively.

By choosing wisely, businesses can protect their IoT systems effectively. Prioritizing data security is key in the IIoT landscape, ensuring compliance and trust while safeguarding sensitive information. Stay updated and secure our digital future together.

5.4   STANDARDS AND PROTOCOLS FOR IIoT COMPLIANCE

In this section, we embark on a comprehensive exploration of the intricate landscape surrounding standards development organizations (SDOs), industry consortia, and associations engaged in the development and adoption of standards. These entities play a pivotal role in shaping the framework within which industries operate, influencing technological innovation, interoperability, and global trade. Understanding the dynamics and interactions among these organizations is essential for stakeholders across various sectors, as they navigate the complexities of standardization processes and leverage standards to drive growth, competitiveness, and societal impact.

5.4.1   STANDARDS DEVELOPMENT ORGANIZATIONS (SDOs)

SDOs play a crucial role in creating and disseminating standards, which are established, repeatable, and documented ways of doing things [25]. These standards are typically recognized and accredited by national or international authorities, with the International Organization for Standardization (ISO) being one of the most well-known examples. SDOs often collaborate with accredited institutions and have defined processes for standardization.

For instance, the adoption of an ISO standard involves reviews and votes by SDOs of member countries, such as the American National Standards Institute (ANSI) for the United States, the Deutsches Institut für Normung (DIN) for Germany, and the Association Française de Normalisation (AFNOR) for France. Additionally, organizations like the Object Management Group (OMG), the parent organization of the IIC, have agreements with ISO that facilitate the submission of adopted specifications as "publicly available specifications" (PAS), expediting the adoption process as an ISO standard. This streamlined process helps ensure the harmonization and global acceptance of standards, benefiting industries and economies worldwide.

5.4.2   INDUSTRY CONSORTIUMS AND ASSOCIATIONS

A consortium is an association of individuals, firms, organizations, or governments with the aim of collaborating on a common activity or pooling resources to achieve shared goals. The IIC and its counterpart, the digital twin consortium (DTC), are examples of such consortia [26]. These groups bring together member companies to address challenges and opportunities in the industrial sector, focusing on defining requirements for new technologies and business models.

While consortiums like the IIC do not typically develop standards themselves, they often support them. When a requirement for a standard arises through collaboration among members, the consortium may identify an existing standard or encourage the development of a new one by engaging with SDOs. Consortium members may contribute as subject matter experts (SME) to define requirements and test newly developed standards.

Consortiums and industry associations serve various purposes, including knowledge sharing, networking, policy influence, and technology development. Some organize industry conferences to facilitate networking and knowledge exchange, while others engage in policy advocacy to influence government regulations. Additionally, some consortia facilitate the development of shared technologies, such as open-source projects. Lastly, some consortiums and associations develop specifications that, although not formal standards, are voluntarily adopted by members, effectively becoming de facto standards.

5.4.3 CATEGORIES OF STANDARDS

External standards encompass four main classifications: open, closed, de jure, and de facto [27].

1. Open standards provide inclusivity in their development process, allowing anyone to participate without barriers or costs. They are publicly available, with minimal restrictions on their use. Open standards are typically developed in public forums with open membership and transparent procedures, aiming for vendor neutrality. Licensing terms such as royalty-free (RF) or reasonable and non-discriminatory (RAND) are common, ensuring accessibility to all stakeholders.
2. Closed standards, in contrast, are developed and licensed by a select few companies or individuals, often excluding input from the broader user community. These standards may have restrictive licensing requirements, limiting their use to members only, or imposing costs for access and usage.
3. De jure standards hold legal endorsement and protection, sanctioned by official standards organizations like ANSI or the Internet Engineering Task Force (IETF). These standards are formally recognized and enforced by law, providing a framework for compliance and interoperability. Examples include ASCII for text files, SCSI for computer interfaces, and TCP/IP for internet communication.
4. De facto standards encompass a wide range of standards, which may be open or closed, and are often established by organizations or individuals with primary goals other than standardization. These standards gain recognition through public acceptance or licensing agreements and can hold the same significance as official de jure standards. Examples include methodology books by renowned authors, documents from requirements organizations like Plattform Industrie 4.0 and the IIC, methodology documents from open-source projects like those from the Linux Foundation or Eclipse Foundation, government agency publications such as the United States Federal Information Processing Standards, and file formats like ".doc" and ".docx" specified by Microsoft. Despite not always being referred to as standards, de facto standards carry substantial weight and influence in various industries and sectors.

Each category of external standards serves distinct purposes and comes with its own set of implications for stakeholders in various industries. Understanding

TABLE 5.2

Comprehensive Overview of Technology Standards in Industrial IoT (IIoT) Environments

Technology Standards	Summary
Operating Systems	Modern IIoT systems utilize standard operating systems like Linux, often optimized for resource-constrained environments.
Cloud Deployment Technologies	Emerging standards for cloud interoperability, such as the open container initiative (OCI) and cloud native computing foundation (CNCF), aim to prevent vendor lock-in in IIoT deployments.
Data Management	New structures like ontologies and knowledge graphs are vital for data analytics and machine learning in IIoT contexts. Standards like RDF, OWL, and SPARQL facilitate data understanding, with ISO 15000-5 offering a methodology for business vocabularies.
Systems Modeling and Interoperability	Systems Modeling Language (SysML) aids in modeling beyond data, while efforts like the Open Industrial Interoperability Ecosystem (OOIE) address challenges in asset-intensive systems in IIoT environments.
Information Exchange	Various standards ensure reliable and secure real-time data communication in IIoT, such as MQTT, OPC/UA, and DDS. Choosing between these protocols requires assessing system requirements specific to IIoT applications.
Security	Security measures like device authentication and data encryption rely on standards such as PKI and CSA recommendations, ensuring secure IIoT deployments. Protocols like DDS-Security enhance security in information exchange for IIoT systems.
Artificial Intelligence (AI) and Machine Learning	Developing standards for AI/ML, such as those by ISO/IEC JTC 1/SC 42 and OMG, aims to ensure AI trustworthiness and address issues like metadata, neural network sensitivity, and ethical AI within IIoT frameworks.

these categories helps organizations navigate the complex landscape of standards adoption and compliance effectively. The various parts of IIC's Industrial Internet Reference Architecture (IIRA) provide useful guidance. Table 5.2 shows a potential (but not exhaustive) checklist of technology areas to consider for IIoT environments.

5.5 ETHICAL CONSIDERATIONS AND RESPONSIBLE IMPLEMENTATION OF AI IN INDUSTRIAL SETTINGS

In today's industrial landscape, the integration of AI into various operational facets has become pervasive. Understanding how AI systems should be designed to operate responsibly while meeting stakeholder expectations and regulatory requirements is paramount. These entail addressing concerns related to reliability, data privacy, transparency, explainability, and ethical considerations. By navigating these ethical dilemmas and integrating responsible AI practices into the development of industrial AI-based solutions, manufacturers not only enhance their credibility but also bolster their reputation in the market. However, achieving this balance is complex, given the heightened requirements of industrial applications compared to AI products intended for consumer markets, including factors like reliability, verifiability, and safety, which demand meticulous attention.

Many industrial enterprises are actively engaging with ethical and environmental concerns surrounding the use of AI within their corporate social responsibility strategies [28]. They are exploring the implementation of "ethical algorithms" to mitigate the risks of unethical behavior in AI systems. Responsible AI design does not have a one-size-fits-all definition, as organizations often develop their own terminologies and methodologies. Nevertheless, at its core, it involves adhering to design principles that allow AI systems to justify and be held accountable for their decisions. In industrial settings, this means facilitating human inspection of AI algorithms and models' functionality, ensuring transparency and ethical operation throughout the system's lifecycle. However, achieving this transparency and ethical operation is a continuous process, demanding ongoing efforts to reassess ethical compliance, particularly in rapidly evolving AI systems and environments where safety-critical considerations prevail.

5.5.1 ESSENTIAL REQUIREMENTS FOR INDUSTRY-GRADE AI SYSTEMS IN IIoT ENVIRONMENTS

In the rapidly evolving landscape of IIoT, the integration of advanced AI systems holds significant promise for enhancing efficiency, productivity, and automation across industrial operations. However, to harness the full potential of AI in industrial settings, it is imperative to establish industry-grade AI systems that meet stringent requirements for reliability, security, privacy, transparency, and accountability. In this context, understanding and defining the essential requirements for industry-grade AI systems becomes paramount. This section explores the key requirements identified by the AI4DI project, delving into the critical aspects that underpin the development and deployment of trustworthy and resilient AI systems within IIoT environments (Table 5.3).

5.5.2 GUIDELINES FOR ETHICAL IMPLEMENTATION OF AI IN INDUSTRIAL SETTINGS

AI is having an expanding impact on all industrial sectors, which has led to the development of multiple sets of ethical standards for AI from professional associations and industry groups. The IEEE Standards Association and the British Standards Institute have just released new ethical recommendations. In order to predict unforeseen or unexpected behaviors, the IEEE emphasizes that researchers should adopt a "safety mindset." It also suggests that social and moral norms should be taken into account while designing AI technologies and applications.

The ubiquity of AI-based industrial process solutions prompts concerns that degree programs connected to AI do not adequately prepare designers with the necessary ethical awareness. Due to common demands and problems around the globe, the industry IEEE introduced the Ethics Certification Program for Autonomous and Intelligent Systems (ECPAIS) [29]. The aim of the program is to establish five fundamental principles to be taken into account in the design and implementation of AI and ethics: adherence to current human rights frameworks, improving human wellbeing, presumably to ensure accountable and responsible design, transparent technology, and the ability to track misuse. These principles aim to advance

TABLE 5.3

Requirements for Industry-Grade IIoT Systems

Requirement	Description
Explainable	IIoT systems must provide explanations for their decisions and actions, allowing stakeholders to understand and track failures effectively.
Available	IIoT applications should ensure high availability to prevent economic losses from system outages. They should also perform autonomously and integrate quickly into new applications and processes.
Trustworthy	As IIoT devices become more interconnected, trustworthiness becomes essential. Systems must have verifiable identities, and vulnerabilities must be promptly reported and addressed.
Secure	IIoT systems must implement robust security measures to withstand various attack vectors, ensuring data integrity and confidentiality. Encryption and authentication mechanisms should secure communication between devices.
Safe	IIoT systems operating in conjunction with humans or controlling safety-critical processes must comply with safety standards to prevent accidents. They should react quickly to unforeseen events and operate with low latency.
Private	IIoT systems must handle sensitive data confidentially, processing it locally at the edge to maintain privacy and confidentiality. Data should be protected from external access to ensure compliance with privacy regulations.
Transparent	IIoT systems should provide transparency in their state, actions, and decisions, facilitated by digital twins and data visualization methods. Stakeholders should be able to inspect and understand system behavior easily.
Fair	IIoT technologies must adhere to fairness and compliance standards, ensuring equitable decision-making processes in accordance with industrial regulations.
Inclusive	IIoT systems should involve humans and existing systems in their operations to prevent isolation within processes, production systems, or supply chains.
Collaborative	IIoT systems must collaborate across distributed AI-enabled sub-systems, IoT nodes, and embedded devices to ensure coherent operation and human-machine collaboration.
Integrative	IIoT systems should be open and flexible, seamlessly integrating into existing systems and processes to facilitate their adoption according to a sustainable roadmap.
Reliable	IIoT systems must be reliable and dependable, ensuring continuous operation with minimal maintenance and system outages, particularly in mission-critical production environments.
Resilient	IIoT systems should remain stable even in the face of process failures and should be capable of detecting and compensating for failures in the future.
Accountable	IIoT systems supporting or replacing human decisions must be accountable for their output, enabling suppliers to be held responsible.
Verifiable	IIoT systems must meet validation, verification, and certification standards for safety, mission-critical, and business-critical tasks, ensuring correctness and reliability in their applications.

transparency, accountability, and reduction in algorithmic bias in autonomous and intelligent systems.

The OECD AI principles, which support trustworthy and innovative AI while upholding democratic ideals and human rights, are another noteworthy project. The OECD Council recommendation on AI was endorsed by OECD member countries in May 2019, marking the adoption of the principles [30].

5.6   LEGAL AND REGULATORY CONSIDERATIONS FOR EDGE COMPUTING IN IIoT SETTINGS

In the rapidly evolving landscape of IIoT, edge computing plays a pivotal role in enhancing efficiency, reducing latency, and enabling real-time decision-making. However, alongside its technological advancements, edge computing introduces various legal and regulatory considerations that must be carefully navigated by organizations operating in IIoT environments.

1. Data Privacy and Security: Edge computing involves processing data closer to its source, often at the device or sensor level, before transmitting it to centralized servers or the cloud. While this decentralized approach offers benefits such as reduced bandwidth usage and faster response times, it also raises concerns regarding data privacy and security. Organizations must adhere to stringent data protection regulations, such as GDPR in Europe or the HIPAA in the United States, to ensure that sensitive information collected at the edge is properly handled, stored, and transmitted. Compliance with these regulations requires implementing robust encryption protocols, access controls, and data anonymization techniques to safeguard against unauthorized access or data breaches.

2. Jurisdictional Compliance: Edge computing introduces complexities regarding jurisdictional compliance, particularly in multinational organizations with operations spanning multiple countries. Different regions may have distinct legal frameworks governing data privacy, cybersecurity, and intellectual property (IP) rights. Navigating these diverse regulations requires a comprehensive understanding of the legal landscape in each jurisdiction where IIoT systems are deployed. Organizations must ensure that their edge computing infrastructure complies with applicable laws and regulations in each jurisdiction, including data residency requirements, cross-border data transfer restrictions, and industry-specific compliance standards.

3. Liability and Accountability: As edge computing enables autonomous decision-making and real-time control within IIoT environments, questions arise regarding liability and accountability in the event of system failures, accidents, or data breaches. Determining responsibility may involve various stakeholders, including device manufacturers, software developers, service providers, and end-users. Organizations must establish clear contractual agreements, service level agreements (SLAs), and liability clauses to define roles, responsibilities, and recourse mechanisms in case of legal disputes or damages arising from edge computing operations. Additionally, obtaining appropriate insurance coverage, such as cyber liability insurance, can help mitigate financial risks associated with potential legal liabilities.

While edge computing offers numerous benefits for IIoT deployments, organizations must carefully address the legal and regulatory implications to ensure compliance, mitigate risks, and uphold trust among stakeholders. By proactively addressing data privacy, jurisdictional compliance, and liability concerns, organizations can leverage the transformative potential of edge computing while navigating the complex legal landscape effectively.

5.6.1 INTELLECTUAL PROPERTY (IP) RIGHTS AND DATA OWNERSHIP

Another relevant aspect to consider in the context of legal and regulatory implications of edge computing in IIoT environments is IP rights and data ownership.

a. IP Rights: Edge computing involves processing data generated by sensors, devices, and machinery at the edge of the network. This data often contains valuable insights, proprietary algorithms, and innovative solutions developed by organizations. As such, protecting IP rights becomes paramount to safeguarding investments in research and development. Organizations must assess the ownership of IP created or utilized within edge computing systems, including algorithms, software code, sensor designs, and data analytics models. Implementing robust IP protection strategies, such as patents, copyrights, trademarks, and trade secrets, can help safeguard against unauthorized use, reproduction, or distribution of proprietary assets.

b. Data Ownership and Control: Edge computing blurs the traditional boundaries between data ownership and control, particularly in collaborative IIoT ecosystems involving multiple stakeholders. Organizations must clarify ownership rights and data usage permissions concerning the data collected, processed, and shared within edge computing architectures. Clear contractual agreements and data sharing agreements should outline the rights and responsibilities of each party regarding data ownership, access rights, usage restrictions, and data monetization opportunities. Additionally, organizations must consider data sovereignty principles, which stipulate that data collected within a specific jurisdiction remains subject to the laws and regulations of that jurisdiction, even when processed or stored elsewhere.

c. Licensing and Compliance: Edge computing solutions often incorporate third-party software components, open-source libraries, and proprietary algorithms licensed from external vendors or developers. Ensuring compliance with licensing agreements, open-source licenses, and IP rights associated with these components is essential to avoid legal disputes, infringement claims, or contractual breaches. Organizations must conduct thorough due diligence to assess the licensing terms, usage restrictions, and redistribution requirements of third-party software incorporated into edge computing solutions. Establishing mechanisms for tracking software dependencies, managing license obligations, and enforcing compliance with licensing agreements can help mitigate legal risks and ensure regulatory compliance.

Addressing IP rights, data ownership, and licensing considerations is essential for organizations deploying edge computing solutions in IIoT environments. By establishing clear ownership rights, protecting IP assets, and ensuring compliance with licensing agreements, organizations can mitigate legal risks, foster innovation, and promote responsible data stewardship in the era of edge computing.

5.7 CONCLUSION

Navigating compliance within the IIoT landscape necessitates a multifaceted approach that addresses organizational needs, regulatory challenges, data security, and ethical considerations. As outlined in this chapter, organizations must first recognize the imperative for IIoT integration and understand the associated compliance challenges, including data security and privacy concerns. By implementing robust compliance frameworks tailored to IIoT deployments, businesses can mitigate risks and ensure adherence to industry standards and protocols. Additionally, prioritizing data security and privacy compliance is essential to safeguarding sensitive information in IIoT systems. Furthermore, adherence to ethical principles and responsible AI implementation practices is crucial for fostering trust and maintaining integrity in industrial settings. Finally, considering the legal and regulatory implications of edge computing in IIoT environments is paramount for ensuring compliance with evolving laws and standards. Overall, by addressing these key areas, organizations can navigate the complexities of compliance in the IIoT landscape effectively, facilitating innovation while upholding legal and ethical standards.

REFERENCES

1. European Commission, Directorate-General for the Information Society and Media, Friess, P., Guillemin, P., & Sundmaeker, H. (2010). Vision and challenges for realising the internet of things. Publications Office.
2. Gubbi, J., Buyya, R., Marusic, S., & Palaniswami, M. (2013). Internet of things (IoT): A vision, architectural elements, and future directions. *Future Generation Computer Systems*, 29(7), 1645–1660.
3. Haller, S., Karnouskos, S., & Schroth, C. (2008). The future of industrial communication: Automation networks in the era of the internet of things and industry 4.0. *IEEE Industrial Electronics Magazine*, 11(1), 17–27.
4. Berman, S. J. (2012). Digital transformation: Opportunities to create new business models. *Strategy & Leadership*, 40(2), 16–24.
5. Zaslavsky, A., Perera, C., & Georgakopoulos, D. (2013). Sensing as a service and big data. In 2013 IEEE World Forum on Internet of Things (WF-IoT) (pp. 287–291). IEEE.
6. Mo, H., & Chen, Z. (2016). The internet of things for ambient assisted living. In E-health and Bioengineering Conference (EHB) (pp. 1–4). IEEE.
7. National Institute of Standards and Technology (NIST). (n.d.). NIST Cybersecurity Framework. Retrieved from https://www.nist.gov/cyberframework.
8. International Organization for Standardization (ISO) and International Electrotechnical Commission (IEC) (2013). "Information Technology — Security Techniques — Information Security Management Systems — Requirements," ISO/IEC 27001.
9. International Electrotechnical Commission (IEC). (n.d.). IEC 62443: Industrial Automation and Control Systems Security. [Online]. Available: [ISA/IEC 62443 Series of Standards - ISA].
10. European Union. (n.d). General Data Protection Regulation (GDPR). Retrieved from https://gdpr-info.eu/
11. U.S. Department of Health & Human Services (2018). "Health Insurance Portability and Accountability Act (HIPAA)," [Online]. Available: https://www.hhs.gov/hipaa/
12. Industrial Internet Consortium (2016). "Industrial Internet Security Framework - Industry IoT Consortium." Available: https://www.iiconsortium.org/iisf/ [Online].

13. Industrial Internet Consortium (2023). "Security Framework (IISF) - iiconsortium. org." Available: https://www.iiconsortium.org/wp-content/uploads/sites/2/2023/06/IISF-Version-2.pdf [Online].
14. Industrial Internet Consortium (2016). "Industrial Internet Security Framework (IISF) - iiconsortium.org." Available: https://hub.iiconsortium.org/iisf [Online].
15. Čolaković, A., & Hadžialić, M. (2018). Internet of things (IoT): A review of enabling technologies, challenges, and open research issues. *Computer Networks*, 144, 17–39.
16. Younan, M., Houssein, E. H., Elhoseny, M., & Ali, A. A. (2020). Challenges and recommended technologies for the industrial internet of things: A comprehensive review. *Measurement*, 151, 107198.
17. Bhutta, M. N. M., Bhattia, S., Alojail, M. A., Nisar, K., Cao, Y., Chaudhry, S. A., & Sun, Z. (2022). Towards secure IoT-based payments by extension of payment card industry data security standard (PCI DSS). *Wireless Communications and Mobile Computing*, 2022, 1–10.
18. Gebremichael, T., Ledwaba, L. P., Eldefrawy, M. H., Hancke, G. P., Pereira, N., Gidlund, M., & Akerberg, J. (2020). Security and privacy in the industrial internet of things: Current standards and future challenges. *IEEE Access*, 8, 152351–152366.
19. Tange, K., De Donno, M., Fafoutis, X., & Dragoni, N. (2020). A systematic survey of industrial internet of things security: Requirements and fog computing opportunities. *IEEE Communications Surveys & Tutorials*, 22(4), 2489–2520.
20. Agrawal, M., Zhou, J., & Chang, D. (2019). A survey on lightweight authenticated encryption and challenges for securing industrial IoT. *Security and Privacy Trends in the Industrial Internet of Things*, 71–94.
21. Esfahani, A., Mantas, G., Matischek, R., Saghezchi, F. B., Rodriguez, J., Bicaku, A.,..., & Bastos, J. (2017). A lightweight authentication mechanism for M2M communications in industrial IoT environment. *IEEE Internet of Things Journal*, 6(1), 288–296.
22. Diego, J. D. D. H., Saldana, J., Fernández-Navajas, J., & Ruiz-Mas, J. (2019). IOTSafe, decoupling security from applications for a safer IoT. *IEEE Access*, 7, 29942–29962.
23. James, E., & Rabbi, F. (2023). Fortifying the IoT landscape: Strategies to counter security risks in connected systems. *Tensorgate Journal of Sustainable Technology and Infrastructure for Developing Countries*, 6(1), 32–46.
24. Zarpelão, B. B., Miani, R. S., Kawakani, C. T., & de Alvarenga, S. C. (2017). A survey of intrusion detection in internet of things. *Journal of Network and Computer Applications*, 84, 25–37.
25. Standards Coordinating Body. (n.d.). Standards Development Organizations (SDOs). Retrieved from https://www.standardscoordinatingbody.org/sdos.
26. Usländer, T., Baumann, M., Boschert, S., Rosen, R., Sauer, O., Stojanovic, L., & Wehrstedt, J. C. (2022). Symbiotic evolution of digital twin systems and dataspaces. *Automation*, 3(3), 378–399.
27. Lou, D., Holler, J., Patel, D., Graf, U., & Gillmore, M. (2021). The Industrial Internet of Things Networking Framework. *Industrial IoT Consortium*.
28. Vermesan, O., De Luca, C., John, R., Coppola, M., Debaillie, B., & Urlini, G. (2022). Ethical Considerations and Trustworthy Industrial AI Systems. *Intelligent Edge-Embedded Technologies for Digitising Industry*.
29. IEEE Standards Association. (2020). The ethics certification program for autonomous and intelligent systems (ECPAIS).
30. Organization for Economic Co-operation and Development (2019). Principles on Artificial Intelligence. Paris: OECD. Available online at: https://legalinstruments.oecd. org/en/instruments/OECD-LEGAL-0449.

6 Lightweight Secure Key Authentication Scheme for Industrial Internet of Things

M. Anoop, Ismail Keshta, Vrince Vimal, Mukesh Soni, and Navruzbek Shavkatov

6.1 INTRODUCTION

In recent years, Industrial Internet of Things (IIoT) technologies [1] based on group key agreements have been widely applied in industries such as manufacturing and finance. Among them, industrial entities generate vast amounts of data closely related to IIoT equipment during production, transportation, and usage. Therefore, constructing a lightweight secure key authentication scheme, efficient, and highly scalable group key agreement protocol to ensure secure communication between different entities has become a hot topic in both academia and industry.

Since the concept of key agreements was first proposed by author [2], research on key agreements has been flourishing in both academia and industry [3]. As a special form of key agreement [4], group key agreement protocols involve group members negotiating group [5] session keys during communication, ensuring secure and efficient communication among group members. However, the key agreement scheme proposed in literature [6] can only generate keys for two parties and cannot verify the identity information of the session key holder. In the face of man-in-the-middle attacks, the security of the protocol cannot be guaranteed. Therefore, author [7], by utilizing the design of symmetric incomplete block designs, proposed a group key agreement algorithm based on block designs, which to some extent solved the problems of weak security and low communication efficiency faced in group key agreements. In order to further reduce the communication overhead of group key agreement protocols, this paper introduces the mathematical structure of incomplete block designs and proposes a structured group key agreement model, making the number of participating members more flexible and the protocol more scalable.

In 1946, author [8] first proposed the concept of pairing operations, which achieved multiplicative homomorphic hiding by mapping elements from two additive groups to a multiplicative group, and it has been widely used in the field of cryptography. However, performing online pairing operations consumes a considerable amount of computational power, which can affect user experience to some extent. Especially when there are a large number of sensors and actuators with low computational power during communication, protocols based on pairing operations are not

    DOI: 10.1201/9781003466284-6

the best choice. Therefore, in order to solve the problem of low computational power and high computational overhead of communication devices in IIoT, this paper proposes an authentication protocol based on elliptic curve Qu-Vanstone (ECQV) [9], which does not require pairing operations, thus effectively addressing the problem of high computational overhead of low-power devices in IIoT. In summary, the innovations of the proposed protocol in this chapter are as follows:

To reduce computational overhead, this paper uses the ECQV authentication protocol to avoid performing pairing operations. Experimental results show that the proposed scheme has significant advantages in terms of computational overhead, with a computational complexity of $On\sqrt{n}\ m$, where n is the number of participating members (including members and volunteers), and m represents the extent of extension of the finite field $G_q{}^m$.

To enhance the security of the protocol, the security of the proposed protocol was proven by utilizing the elliptic curve decisional Diffie–Hellman (ECDDH) assumption. Security analysis shows that all members contribute equally to the group session key, ensuring the security of group data sharing in IIoT.

In order to reduce the communication overhead of the protocol and improve its scalability, this paper extends existing group key agreement protocols using asymmetric incomplete block designs. The supported number of members is extended from q^2 to $q^2 + q + 1$. Specifically, this paper constructs a restructured asymmetric block design structure and applies asymmetric block design to group key agreements for the first time. Through this group key agreement protocol based on asymmetric block design, the supported total number of members is extended from $q^2 + q + 1$ to q^2 and $q^2 + q + 1$, where q is a prime number. Compared with protocols based on symmetric block designs [7], protocols based on asymmetric block designs significantly reduce communication overhead and further enhance scalability.

Rest of the organization of the chapter goes as follows: Section 6.2 describes literature review on security models, information interaction, and group key agreements. In Section 6.3, foundational concepts include cryptographic primitives and group dynamics. In Section 6.4, threats and system architecture are presented, and in Section 6.5, organizational data transfer details, and in Section 6.6, How to establish an agreement using a shared group key are shown. In Section 6.7, formal analysis proves protocol security against typical attackers, and in Section 6.8, evidence-based performance and scalability assessment, and also in Section 6.9, results, importance, and future research are presented.

6.2   RELATED WORK

Traditional structured group key agreement. In the IIoT environment, with the increasing demand for online collaboration and the expansion of sharing platforms, traditional two- or three-party key agreement schemes can no longer meet practical requirements, necessitating the urgent proposal of a new group key agreement scheme. Author [10] first designed a ring-based interaction model supporting key agreements and expanded the number of negotiators to multiple parties. However, this protocol can only resist passive attacks. A heuristic security proof and a group key agreement mechanism based on binary trees were presented by the author [11] to accommodate the ever-changing needs of group users. After that, a ternary tree-based group key agreement technique was created by author [12]. Unfortunately, it is still unable to confirm the members' identities, even after expanding the binary tree scheme. A non-authenticated group key agreement mechanism was suggested by the author [13]. This protocol can withstand spoofing attacks launched by external nodes and completes information exchange across groups in just two rounds of communication. One negative aspect is that it can't withstand internal attacks that are hostile.

A group key agreement system based on passwords was suggested by author [14] as a solution to this problem. There is a linear relationship between the number of communication rounds and the number of participating members, even though this protocol can verify the user's identity. Invented in 2023 by author [15], this technique for group key agreement only requires one round.

With this protocol, users cannot deny the messages they send, but when the number of users is n, the communication overhead reaches $O(n^2)$.

Lightweight secure key authentication scheme is based on combinatorial structures. In order to further ensure the security of group data sharing, researchers have designed a highly lightweight secure key authentication scheme agreement protocol. Author [16] used the combinatorial structure of Latin squares to construct a group key agreement protocol based on Latin squares. This scheme introduces combinatorial structures into group key agreement, which to some extent reduces communication overhead. However, due to the excessively high number of communication rounds, communication efficiency is low. Therefore, author [17] proposed an identity-based key agreement protocol by introducing the mathematical structure of block designs to achieve entity authentication for group key agreements. However, this protocol supports communication among a maximum of seven members, lacking some flexibility. Subsequently, author [7] optimized the data structure of block designs and proposed a group key agreement protocol based on block designs, which, with the inclusion of volunteers, can support communication among up to n members and compress communication overhead to $On\sqrt{n}$. However, since this protocol introduces Weil pairing operations, it leads to excessive computational overhead.

Group key agreement based on non-pairing operation techniques. Author [18] created a vehicle authentication and key agreement method for real-time wireless data transfer in vehicular networks. Given long-term key non-disclosure, this method can also secure session keys from short-term key leakage. However, when the number of vehicles and cloud servers is n and m, respectively, the communication overhead reaches $O(nm)$. In 2022, author [19] proposed a public key group key agreement protocol that does not require pairing operations, significantly reducing the computational overhead caused by pairing operations. However, since this protocol uses broadcast communication, it also increases communication overhead.

In summary, existing group key agreement technologies mainly suffer from the following shortcomings:

Although traditional structured group key agreement can expand the protocol from two or three parties to multiple parties, it faces challenges of low security and high communication rounds.

Although group key agreement based on combinatorial structures can effectively reduce communication rounds, it incurs high computational and communication overheads and has low scalability.

Although group key agreement based on non-pairing operation techniques has lower computational overhead, it suffers from high communication overhead.

6.3 BACKGROUND KNOWLEDGE

This chapter first introduces security assumptions and the cryptographic operations involved; then presents the ECQV lightweight secure key authentication scheme; finally, provides a detailed introduction to block designs. The parameters and their specific meanings involved in this paper are listed in Table 6.1.

6.3.1 SECURITY ASSUMPTIONS AND CRYPTOGRAPHIC OPERATIONS

In public-key cryptography, the security of protocols is of paramount importance. Attackers and attack targets can be detailed using security models, and specific security proofs are based on the ECDDH assumption, defined as follows:

Definition 1: Let F_q be an elliptic curve over the finite field F_q, and H be a point of order n on F_q. X, Y, and A are three points on F_q, satisfying $X = xH, Y = yH$, where x and y are two randomly chosen integers from G_q, and A is a randomly selected point on Ep. Given H, X, Y, ECDD H assumption requires determining whether A equals xyH through the algorithm Algo. If they are equal, output 1; otherwise, output 0. The advantage ε of the algorithm Algo in solving the ECDDH problem is defined as follows:

$$|\Pr[\text{Algo}(H, X, Y, A) = 1]$$
$$\Pr[\text{Algo}(H, X, Y, xyH) = 1]| \geq \varepsilon$$

According to [20], the ECDDH assumption is violated if the method Algo provides a non-negligible advantage ε in solving the ECDDH issue.

TABLE 6.1
Notation

Parameter	Meaning
w	All the elements that make up the set W
c	Quantity of blocks in total
s	The total number of blocks that have a specific element
k	The variety of components included in every block
λ	Total number of blocks when two items are present
MOD(b,m)	Remainder of dividing b by m
m|n	m divides n
|B|	The number of elements in set B
b≠c(modp)	b and c mod p are not congruent
b^{-1}	The number theory reciprocal of b

6.3.2 CRYPTOGRAPHIC OPERATIONS

To increase protocol security, this work employs a collision-proof single-direction hash function H and symmetric encryption-decryption functions (Enc, Dec) [21]. Using a symmetric key K, plaintext N is encrypted into ciphertext D and ciphertext D is decrypted into plaintext N, expressed as $D = Fnc_K(N)$ and $N = Dec_K(D)$ [22].

6.3.3 ELLIPTIC CURVE QU-VANSTONE (ECQV) AUTHENTICATION

ECQV is a lightweight secure key authentication scheme based on elliptic curve cryptography [23], characterized by low computational overhead and high security, making it suitable for IIoT group communication [24]. In this scheme, eavesdroppers cannot obtain the lightweight secure key authentication scheme, through the transmission channel [25]. The implementation steps of ECQV are shown in Algorithm 1.

Algorithm 1: ECQV Authentication

Input: Member i identity ID_i, timestamp u_i.
Output: Success of member i authentication

1. Member i randomly chooses s_i, calculates $S_i = s_i \mathbf{H}$;
2. Member i sends (ID_i, S_i) to the authentication center;
3. The authentication center randomly selects d_e, s_e, calculates $Q_e = d_e \mathbf{H}$, $S_e = s_e \mathbf{H}_7$;
4. The authentication center calculates $\gamma_i = R_i + R_e$, Cesu $i = \text{Encode}(\gamma_i, ID_i, *)$, $a_i = H(\text{Cesu}_i, u_i)s_c + d_c$;
5. The authentication center sends $(R_c, ai_{ii}, \text{Cesu}_i, u_i)$ to member i_4;
6. Member i calculates $\mathbf{R}_i = H(\text{Ceru}_i, \mathbf{u}_i)\gamma_i + \mathbf{R}_c$, $d_i = H(\text{Ceru}_i, \mathbf{u}_i)r_i + a_i$;
7. If $\mathbf{Q}_i == d_i \mathbf{H}$ then
8. Member i authentication succeeds;
9. Else
10. Member i authentication fails

6.3.4 BLOCK DESIGNS

In combinatorial designs, block designs consist of sets and their subsets, where the subsets in a cluster of subsets are called blocks. The elements within blocks satisfy the overall structural balance. The definition of block designs is as follows:

Definition 2: Let W be a set of cardinality w, i.e., $W = \{1, 2,..., w\}$; $C = \{C_1, C_2,..., C_c\}$ is a set containing c blocks, where C_i a subset of is W, and satisfies $|C_i| = k$, $i = 1, 2, ..., c$. If $\sigma = (W, C)$ satisfies the following conditions, then it is called (w, c, s, k, λ) balanced incomplete block design (BIBD).

1. Every component looks precisely s times.
2. For any set of two components, there is an exact λ times simultaneously. If conditions 1 and 2 are met, then σ is called a BIBD.
3. Parameters k and w of σ must satisfy $k < w$, meaning no block can contain all elements.

4. Parameters d and w of σ must satisfy $d \geq w$. If $d = w$, then σ is called a symmetric block design; otherwise, it is called an asymmetric block design.

For (w, d, s, k, λ) BIBD, it must satisfy $wk = ds$ and $(w-1) = s(k-1)$. If $d = w$ and $k = s$ hold, it is called a symmetric BIBD; otherwise, it is called an asymmetric BIBD.

This chapter constructs a group information sharing model by constructing $(w_1, q, 1)$ and $(w_2, q+1, 1)$ BIBDs, where q is a prime number and $\lambda = 1$. Since $wk = ds$ and $\lambda(w-1) = s(k-1)$, the parameters of the five block designs depend entirely on three of them. Here, (w, k, λ) is chosen as the three main parameters.

In the proposed protocol of this paper, based on the structure of $(w_1, q, 1)$ or $(w_2, q+1, 1)$ BIBDs, each member can determine the sender of the expected message.

6.4 SYSTEM MODEL AND ADVERSARY MODEL

This part will begin by introducing the protocol's system model, which will help to better comprehend the protocol described in this work, and then describe the adversary model, and finally summarize the capabilities of active attackers. The architecture of the system is illustrated in Figure 6.1.

6.4.1 Architecture of the System

The Architecture of the System in this chapter consists of three parts: users, service providers, and authentication centers. Users communicate with each other to obtain group session keys; service providers are responsible for registering users

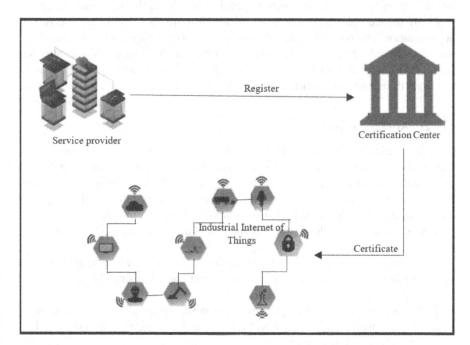

FIGURE 6.1 Architecture of the system.

and authentication centers to ensure that the entities in the protocol are legitimate; authentication centers initialize key materials for users and grant certificates to verify the identity of users. The following will provide detailed explanations for each:

Users

In this protocol, considering the presence of malicious users who do not comply with the rules of group communication, the goal of such users is to make the lightweight secure key authentication scheme of different members, thereby undermining the security and reliability of group communication. According to this protocol, during the group key negotiation protocols, group members prevent illegal situations by checking the identities of members.

Service Providers

Service providers are honest but curious entities. According to this protocol, since service providers cannot obtain the private keys of other members or the group's private key, they cannot access the content of group communication.

Authentication Centers

Authentication centers are fully trusted entities responsible for building key materials for each member and issuing digital certificates to verify the legitimacy of user identity. Although it is impossible to compute the private keys of any member, the authentication center can compute the lightweight secure key authentication scheme. Furthermore, the authentication center cannot impersonate any member to ensure the security and reliability of the protocol. In the IIoT, authentication centers can be operated by large institutions with certificate distribution qualifications. Since the authentication center is a fully trusted entity, channels associated with it are also lightweight secure key authentication schemes. Once the lightweight secure key authentication scheme channel is established, it becomes resistant to man-in-the-middle attacks. With the introduction of timestamps, any attempts by a man-in-the-middle attacker would require more time than the actual communication delay, thus allowing the detection of such attacks.

6.4.2 ADVERSARY MODEL

The communication channel is open, and information faces both passive and active attacks during communication. Passive attacks refer to attackers eavesdropping on the communication channel to gain knowledge of the lightweight secure key authentication scheme, while active attacks involve attackers attempting to disrupt group communication by interrupting sessions or impersonating members.

The adversary model defines the capabilities of active attackers, including:

1. An attacker can get access to the group session key by impersonating a member and communicating with other participants using the member's long-term private key.
2. The adversary can obtain previous session keys and acquire new member key information. Consequently, they can impersonate new members and others for communication.

3. The adversary can obtain a member's private key. Subsequently, they attempt to compute previous group session keys to access previous communication content.

6.5　GROUP INFORMATION INTERACTION MODEL

Suppose there are M members, and let q_n denote the prime number sequence, such as $q_1 = 2$, $q_2 = 3$. Considering the finiteness of M, there must exist a positive integer i_0 such that $q_{i_0-1}^2 + q_{i_0-1} + 1 < M \leq q_{i_0}^2$ or $q_{i_0-1}^2 < M \leq q_{i_0}^2 + q_{i_0} + 1$. For the former case, this chapter adopts the $(w_1, q, 1)$ design and demonstrates the construction and reconstruction phases. For the latter case, this paper applies the $(w_2, q+1, 1)$ design, and the construction and reconstruction phases are presented in [7].

6.5.1　CONSTRUCTION OF $(w_1, q, 1)$ BLOCK DESIGN

Algorithm 2: Generating $(w_1, q, 1)$ Design

Input: Prime number q
Output: $(w_1, q, 1)$ block design C

1. for i = 1, i ≤ q; i++ do
2. for j = 1; j ≤ q; j++ do
3. $C_{i,j} = (i-1)q + j$;
4. end for
5. end for
6. for i = q + 1, i ≤ 2q; i++ do
7. for j = 1, j ≤ q, j++ do
8. $C_{i,j} = (j-1)q + i - q$;
9. end for
10. end for
11. for l = 1; 1 ≤ q − 1; l++ do
12. for i = (l+1)q + 1, i ≤ (l+2)q; i++ do
13. for $j = 1, j \leq q + j$++ do
14. $C_{i,j} = q \text{MOD}((j-i)l, q) + j$
15. end for
16. end for
17. end for

Inspired by reference [21], Algorithm 2 constructs a $(w_1, q, 1)$ group design. The specific implementation process is as follows: Firstly, determine a prime number q_0 based on the number of members. Then, calculate the parameters in this protocol: $w_1 = q^2$, $c_1 = q^2 + q$, $r_1 = q + 1$, $k_1 = q$, $\lambda_1 = 1$.

According to **Definition** 2, $V = 1, 2, ..., w$, where v represents the number of members, and $C = \{C_1, C_2, ..., C_c\}$ represents b groups, each consisting of k members. $C_{i,j}$ denotes the j-th member in the i-th block Bi. All groups form a matrix

C of size $c \times k$, where each row represents $C_{i,j}$ for $j = 1, 2, \ldots, k$, and each column represents $C_{i,j}$ for $i = 1, 2, \ldots, c$. In summary, Algorithm 2 provides the construction method for the $(\tau_1, q, 1)$ group design.

Taking prime number $q = 3$ and selecting the target element in the 9th row and 2nd column as an example, for the $(w_1, q, 1)$ design, compute $C_{Q.2}$.

$$C_{9,2} = q \cdot \text{MOD}((j-i)l, q) + j$$
$$= 3 \cdot \text{MOD}(-1, 3) + 2 = 3 \times 2 + 2 = 8 \tag{6.1}$$

It yields member 8 as the second member of block C_9.

Definition 3: In the $(w_1, q, 1)$-group design model, define T_x as a continuous set containing all elements of groups, calculated as $T_x = \{C_{\ell(x-1)+1}, C_{q(x-1)+2}, \ldots, C_{qx}\}$, $x = 1, 2, \ldots, q+1$.

6.5.2 RECONSTRUCTING THE $(w_1, q, 1)$-GROUP DESIGN

After constructing the original $(w_1, q, 1)$-group design, to generate group session keys, the group design structure must satisfy the property that group Bi contains members i, $i = 1, 2, \ldots, q^2$. Hence, this paper reconstructs the $(w_1, q, 1)$-group design.

Definition 4: Let $I_x = \{i : x \in C_i\}$ be the index set containing member x in the group. For example, $I_1 = \{1, q+1, 2q+1, \ldots, q^2 - q + 1\}$. Obviously, element 1 appears in $I_1, I_2 , \ldots, I_q$. Denote the reconstructed $(w_1, q, 1)$-group design group as E_i, $1 \le i \le q^2 + q$.

Lemma 1: The general formula for I_x is as follows. $I_{(i_0-1)q+j_0} = \{i_0, q+j_0, (l+1)q+1+ \text{MOD}(-l^{-1}(i_0-1)+j_0-1, q), l = 1, 2, \ldots, q-2\}$ Where $1 \le i_0, j_0 \le q$.

Proof: According to Algorithm 2, note that in T_1, $C_{i,j} = (i-1)q+j$, $1 \le i$, $j \le q_0$. Solving the equation $C_{i,j} = x = (i_0-1)q + j_0$ we get the row index for element x as i_0. In T_2, solving the $C_{i,j} = (j-1)q+i-q = x = (i_0-1)q+j_0$, $q+1 \le i \le 2q$, $1 \le j \le q$, the solution is $q+j_0$. For $T_k (k \ge 3)$, solving the equation $q\text{MOD}((j-i)l, q)+j = x = (i_0-1)q+j_0$, $(l+1)q+1 \le i \le (l+2)q$, $1 \le j \le q$, the solution is $(l+1)$ $q+1+\text{MOD}(-l^{-1}(i_0-1)+j_0-1, q)$, where $l = 1, 2, \ldots, q-2$. Thus, Lemma 1 is proven.

Lemma 2: $2q_{(i_0-1)q+j_q} = I_{(i_8-1)q+j_q,1+\text{MOD}(i_q+j_q-2,q)}$, and for $1 \le i_0, j \le q$, $r(i_0-1)q+0$ are mutually distinct, forming a permutation of $1, 2, \ldots, q^2$.

Algorithm 3: Reconstructing the $(w_1, q, 1)$ design

Input: Prime number q, $(w_1, q, 1)$-group design C
Output: Reconstructed $(w_1, q, 1)$-group design F

    1. for i = 1; 1 ≤ i ≤ q; i++ do
    2. for $j = 1$, $1 \le j \le q$; j++ do
    3. $F_{(i-1)q+j} = C_{r_i-1}$
    4. End for
    5. End for

Proof: For ease of understanding, let matrix I aggregate all I_x, $1 \le x \le q^2$, column-wise. It is noteworthy that $(k-1)q+1 \le I_{x,k} \le kq$, thus all elements are within the

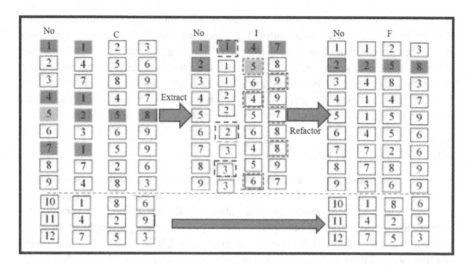

FIGURE 6.2 Schematic diagram of asymmetric balanced incomplete block design (BIBD).

range of 1 to q^2. By fixing i_0, consider q numbers $(i_0 - 1)q + 1$, $(i_0 - 1)q + 2$, ..., $i_0 q$, generally having the same i_0 but different j_0. According to **Algorithm** 3, they belong to different columns of I. Therefore, $q_{(i_0-1)q+j_0}$, $1 \le j_0 \le q$ are distinct from each other.

Hence, it suffices to prove that qx in column $1 + \text{MOD}(i_0 + j_0 - 2, q)$, q_x, $x = (i_0 - 1)q + j_0$ are all distinct. For $u = 1 + MOD(i_0 + j_0 - 2, q)$, this paper will discuss three cases. It is noteworthy that for each u, $1 \le u \le q$, there are exactly p pairs of (i_0, j_0) satisfying $u = 1 + \text{MOD}(i_0 + j_0 - 2, q)$.

Case 1: When $u = 1$, as i_0 ranges from 1 to q, $r_{(i_0-1),+j_0} = i_0$ are all distinct.

Case 2: When $u = 2$, as j_0 ranges from 1 to q, $r_{(i_0-1)q+j_0} = q + j_0$ are all distinct.

Case 3: When $u \ge 3$, as i_0 ranges from 1 to q, it is required to prove that $(u-1)q + 1 + MOD(-(u-2)^{-1}(i_0 - 1) + j_0 - 1, q)$ are all distinct. Considering $u = 1 + MOD(i_0 + j_0 - 2, q)$ fixed, it suffices to prove $(u-2)^{-1}(i_0 - 1) + u - i_0$, $1 \le i_0 \le q$ are all distinct, which is equivalent to $(1 + (u-2)^{-1})i_0$, $1 \le i_0 \le q$ being all distinct. Based on number theory knowledge, the proposition is equivalent to $1 + (u-2)^{-1} \ne 0 (\text{mod} q)$. Solving the inequality, we get $u \ne 1(\text{mod} q)$. Since $u \ge 3$, Case 3 is proven.

In conclusion, Lemma 2 is proven with Figure 6.2.

Lemma 2 ensures the correctness of Algorithm 3. Based on Algorithm 3, this paper accomplishes the reconstruction of $(w_1, q, 1)$ area designs. For example, when $(q = 3)$, $F_2 = C_5 = \{2, 5, 8\}$.

6.6 AGREEMENT ON GROUP KEY PROTOCOL

In this protocol, the agreement on group key process adopts a public key, agreement on group key protocol without the need for pairing operations, including six main stages: in the registration stage, the service provider requests the allocation of the required key materials from the authentication center; in the key initialization stage, the authentication center selects system parameters and prepares key materials for members; and in the group key agreement stage, members receive messages from other members through two rounds of communication. Each member uses the key materials obtained in the key initialization stage to generate a lightweight secure key authentication scheme; in the encryption stage, users within the group can encrypt plaintext using the group public key; in the decryption stage, each user can decrypt cipher text sent by the sender; and in the group update stage, this paper will introduce how the service provider responds to the joining of new members or the departure of old members.

6.6.1   REGISTRATION STAGE

This protocol uses an elliptic curve with generator H and hash function J. The authentication center selects the key dc and computes the corresponding public key $R_d = e_d H$ these system parameters and public key R_d are public. The service provider holds the identity and public key (ID_i, R_i) of members participating in group communication, which can be computed through ECQV to ensure the certification of public keys. The private key of each member is e_i, satisfying $e_i H = R_i$. The service provider sends a request to the authentication center containing the group members' information (ID_i, R_i) to initiate the key initialization stage.

6.6.2   KEY INITIALIZATION STAGE

Upon receiving the request from the service provider, the authentication center selects three system parameters s_c, β_1, β_2, and their corresponding elliptic curve points $R_c = r_c H$, $H_1 = \beta_1 H$, $H_2 = \beta_2 H_o$. Then, the authentication center begins the ECQV process for member i, where R_i replaces R_i, $R_{c,i} = r_{c,i} H$ replaces the elliptic curve random point chosen by the authentication center, and the authentication center calculate $s\gamma_i = R_{c,i} + R_i$. Based on this, the certificate for ID_i is defined as Cert $_i =$ Encode $(\gamma_i, ID_i, *)$. then the authentication center records the timestamp u_i and computes $b_i = J(\text{Cert}_i, u_i) s_{c,i} + e_c$. Additionally, the authentication center prepares a set of elliptic curve ElGamal parameters for encrypting intermediate keys. Specifically, the authentication center selects two random values y_i, k_i for member i and publicly announces $Y_i = y_i H$ and k_i for all group members.

Next, the authentication center computes the key pairs $(tk_{i,1}, k_{i,2})$ for members participating in group communication. The first parameter $tk_{i,1}$ equals $e_c - u_i$, where u_i a randomly chosen integer is. $tk_{i,2}$ is an element pair $(s_{i,1}, s_{i,2})$ satisfying Equation 6.2:

$$u_i = s_{i,1} \beta_1 + s_{i,2} \beta_2 - s_c \tag{6.2}$$

The key pair must be encrypted with its new private key $R_{i,n}$ for sharing with member secrets. Specifically, by encrypting with symmetric private key $ntk_i = J(s_c R_{i,n})$, ciphertext $D_i = Enc_{=k_i}(tk_{i,1}, tk_{i,2})$ is generated.

Then, the message $(ID_i, b_i, R_{e,i}, C_i, R_x, H_1, H_2, u_i)$ is separately sent to each selected member. Upon receiving the message, the member records the current timestamp u_i', checks if the message was sent within the time limit Δu by subtracting the current time from the sending time, and is able to generate their certificate Cert $_i =$ Encode$(S_{d,i} + R_i, ID_i, *)$ and private key $e_{i,n} = J(\text{Cert}_i, t_i) e_i + b_i$ 和 and corresponding public key $R_{i,n} = e_{i,n}H_0$. Note that $s_c R_{i,\pi} = s_c e_{i,n}H = S_c e_{i,\pi}$. Therefore, if member i can recover the key pair $(rk_{i,1}, rk_{i,2})$ using the symmetric key $nsk_i = H(S_d e_{i,n})$, they can confirm the legitimacy of the key material and public key by checking Equation 6.3:

$$R_d + S_d = rk_{i,1}H + r_{i,1}H_1 + r_{i,2}H_2 \tag{6.3}$$

Parameters $S_d, H_1, H_2, S_{d,i}, t_i, i = 1, 2, \ldots, w$ are also obtained by the service provider. Therefore, the service provider can generate $R_{i,n} = H$ (Encode ($S_{d,i} + R_i$, ID_i, *), t_i)($S_{d,i} + R_i$) + R_d. However, since there is no e_i, the service provider cannot obtain the private key $e_{i,n}$. Additionally, the group identity is uniquely defined by the elliptic curve point S_d.

6.6.3 AGREEMENTON GROUP KEY

In the group key sharing phase, users need to generate a group session key through two rounds of communication, specifically using the reconstructed $(\tau_1, q, 1)$ and $(w_2, q+1, 1)$ designs.

Firstly, for each member, two random integers z_i, l_i are selected, where z_i ads randomness to the system and li represents the contribution made by the member to generating the group key. Six parameters $(D_{i,1}), (D_{i,2}), (D_{i:3}), (D_{i,4}), (D_{i,5})$, and $(D_{i,6})$ are constructed, and their specific solutions are as follows:

$$\begin{aligned}
D_{i,1} &= J(w_i R_d) + l_i \\
D_{i,2} &= w_i H \\
D_{i,3} &= w_i H_1 = w_i \beta_1 H \\
D_{i,4} &= w_i H_2 = w_i \beta_2 H \\
D_{i,5} &= w_i S_d = w_i s_d H \\
D_{i,5} &= l_i H
\end{aligned} \tag{6.4}$$

For the message $(D_{i,1}), (D_{i,2}), (D_{i:3}), (D_{i,4}), (D_{i,5})$, and $(D_{i,6})$, member $s_i = w_i - h_i e_{i,n}$ for hashing to generate signature t_i, where $j_i = J(D_{i,1}), (D_{i,2}), (D_{i:3}), (D_{i,4})$, $(D_{i,5})$, and $(u_i)))$, u_i is the timestamp used for identity verification. Employing the group data sharing model in this paper, the actual number of members N corresponds to two cases: $(w_1, q, 1)$ and $(w_1, q+1, 1)$. The following will analyze these two cases separately with the help of Figure 6.3 with Table 6.2.

Case 1: If there exists i_0 such that M satisfies $q_{i_0-1}^2 + q_{i_0-1} + 1 < M \le q_{i_0}^2$, this paper adopts the $(w_1, q_{i_0}, 1)$ region design. Therefore, this paper requires M members and $q_{i_8}^2 - M$ volunteers to participate in group communication, where $w = q_{i_0}^2$ represents the total number of members.

First Round: If $j \in F_i$, $1 \le i, j \le q_{i,0}^2$ or members i and j are in the same region F_x, $q_{i_0}^2 + 1 \le x \le q_{i_8}(q_{i_8} + 1)$, member j sends message $E_j = (ID_j, \text{Cert}_j, D_{j,1}, D_{j,2}, D_{j,3}, D_{j,4}, D_{j,5}, D_{j,6}, t_j, u_j)$ to member i. Since the reconstructed $(w_1, q, 1)$ region design has the property that region F_i contains member i, member i cannot communicate with itself. Therefore, if $j \in F_i(j \ne i)$ or $i, j \in F_x$, $q_{i_0}^2 + 1 \le x \le q_{i_0}(q_{i_0} + 1)$, after member i receives a message from member j, it checks if the message was sent within

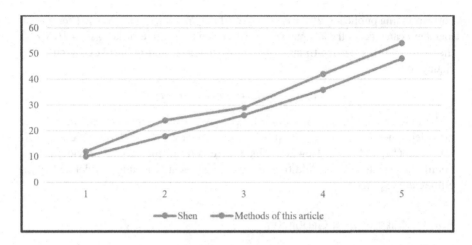

FIGURE 6.3 Comparison of improved membership and user counts.

the time limit Δu [26]. Member i receives messages from $q_{i_0} - 1$ members. Member i calculates Equation (6.5) using the key material $(tk_{i,1}, tk_{i,2})$.

$$
\begin{aligned}
tk_{i,1}D_{j,2} &+ r_{i,1}D_{j,3} + r_{i,2}D_{j,4} - D_{j,5} \\
&= (tk_{i,1} + r_{i,1}\beta_1 + r_{i,2}\beta_2 - r_c) \times w_j G \\
&= d_c w_j G \\
&= w_j Q_c
\end{aligned}
\tag{6.5}
$$

Where:

$$
\begin{aligned}
tk_{i,1} &= e_c - u_i \\
&= e_c - r_{i,1}\beta_1 - r_{i,2}\beta_2 + r_c
\end{aligned}
\tag{6.6}
$$

Member i then calculates l_j using Equation (6.7) and verifies l_j using Equation (6.8).

$$
l_j = D_{j,1} - J(w_j R_c)
\tag{6.7}
$$

$$
D_{j,6} = l_j H
\tag{6.8}
$$

TABLE 6.2

Comparison of Improved Membership and User Counts

Shen	Methods of this Article
10	12
18	24
26	29
36	42
48	54

Additionally [27], member i computes $(T_{i,j,1}, T_{i,j,2}, E'_{i,j})$ through Equation (6.9) using the elliptic curve ElGamal scheme [28], in the second round of communication [29], user j generates the group session key through $(T_{i,j,1}, T_{i,j,2}, E'_{j,i})$.

$$N_{i,j} = \sum_{\substack{x \in F, \backslash (j) \\ 1\sigma_1, j < q_i^2}} l_x$$

$$E'_{i,j} = \bigcup_{\substack{x \in E_1 \backslash (j) \\ 1\sigma_i, j \varepsilon'_j,}} E_x \qquad (6.9)$$

$$T_{i,j+1} = k_j G$$

Second Round: If $i \in E_j$, $1 \le i, j \le q_{i_0}^2$ user i receives a message $(T_{j,i,1}, T_{j,i,2}, E'_{j,i})$ from user j, decrypts it using Equation (6.10), checks if the message was sent within the time limit Δu to resist replay attacks.

$$T_{j,i,2} - y_i T_{j,i,1} = N_{j,i} \qquad (6.10)$$

In fact, each $(T_{j,i,1}, T_{j,i,2})$ contributes $q_{i_0} - 1$ messages to generating the group session key. For T_x, $1 \le x \le q_{i_0}$, user i receives $q_{i_0}(q_{i_0} - 1)$ messages. For $T_{q_{i_q}+1}$, user i receives $q_{i_0} - 1$ messages. Using message $E'_{j,i}$, after decrypting $(\gamma_j, ID_j, *) = \text{Decode}(\text{Cert}_j)$, member i completes the verification of user identity through Equation (6.11).

$$\left(\sum_{j, j \ne i} t_j \right) H = \sum_{j, j \ne i} \left(D_{j,2} - g_j G\left(\text{Cert}_j, u_j \right) \gamma_j \right) - \left(\sum_{j, j \ne i} g_j \right) R_d \qquad (6.11)$$

Among them, $g_i = G(D_{i,1}, D_{i,2}, D_{i,3}, D_{i,4}, D_{i,5}, D_{i,6}, u_i)$, the integrity of the protocol is verified by aggregating the contributions of all members. The specific calculation formula is shown in Equation (6.12):

$$M = l_i + \sum_{\substack{i \in F_j \\ 1 \le i, j \le q_{i_0}^2}} \mathcal{M}_{j,i} + \sum_{\substack{i, j \in F_k, j \ne i \\ q_0^2 + 1 \le k \le q_{i_0}^2 + q_{i_9}}} l_j$$

$$= l_i + \sum_{\substack{i \in F_j \\ 1 \le i, j \le q_i^2}} \sum_{\{x \in F_j \backslash i\}} l_x + \sum_{\substack{i, j \in F_b, j \ne i \\ q_0^2 + 1 \le k \le g_0 + q_i}} l_j \qquad (6.12)$$

$$= \sum_{i=1}^{q_i^2} l_i$$

Legitimate members can recover l_i from E_i using the key material $(tk_{i,1}, tk_{i,2})$ through Equations (6.4) and (6.5). Therefore, the group session public key is

$(R_d, H_1, H_2, S_c + MH)$, and the private key is the element pair $(tk'_{i,1}, tk_{i,2})$, where $tk'_{i,1}$ is updated to $tk_{i,1} + \mathcal{M}_0$.

Case 2: If there exists i_0 such that N satisfies $q_{i_0-1}^2 < N \le q_{i_0}^2 + q_{i_0} + 1$, a design $(w_2, q_{i_0} + 1, 1$ regions is adopted, requiring N members and $q_{i_0}^2 + q_{i_8} + 1 - N$ volunteers, where w is the total number of members, calculated as $w = q_{i_v}^2 + q_{i_v} + 1$. Similar to Case 1, generating the group session key requires two rounds of communication.

First round: $j \in F_i, 1 \le i, j \le q_{ij}^2 + q_{i_i} + 1$, member j sends message $E_j = (ID_j,$ $Cert_j, D_{j,1}, D_{j,2}, D_{j,3}, D_{j,4}, D_{j,5}, D_{j,6}, t_j, u_j)$ to member i. Although the reconstructed $(w_2, q + 1, 1)$ region possesses properties of member i in region E_i, it cannot communicate internally. If $j \epsilon E_j (j \ne 1)$, then member i receives messages from member j. After receiving messages from q_{i_i} members, member i verifies if the message was sent within the time limit Δt. Similarly, by using their own key material $(tk_{i,1}, tk_{i,2})$ and calculating Equation (6.5), member i recovers the integer l_j from $D_{j,1}$ and verifies it with $D_{i,6}$. Additionally, member i calculates $(T_{i,j,1}, T_{i,j+2}, E'_{i+j})$ according to Equation (6.10) to generate the group session key for the second round member j.

$$M_{i,j} = \sum_{\substack{x \in F_i(j) \\ 1 \le i, j \le q_{i0}^2 + q_{i0} + 1}} l_x$$

$$E'_{i,j} = \bigcup_{\substack{x \in F_i \setminus j\} \\ 1 \le i, j \le q_{i,2}^2 + q_{i_e}}} E_x \qquad (6.13)$$

$$T_{i,j,1} = k_j H$$

$$T_{i,j,2} = M_{i,j} + k_j Y_j$$

Second round: If $i \in F_j, 1 \le i, j \le q_{i0}^2 + q_{i0} + 1$, member i receives $(T_{j,i,1}, T_{j,i,2}, E'_{j,i})$ from member j, decrypts and verifies the message using Equation (6.10), and checks if it was sent within the time limit Δu.

During the aforementioned group session key generation process, each $M_{j,i}$ contributes q_{i_0} messages. For $T_x, 1 \le x \le q_{i_q} + 1$, member i receives $q_{i_0}(q_{i_q} + 1)$ messages. Next, all messages are aggregated and identity verification is performed using Equation (6.11).

The correctness of the protocol is ensured by Equation (6.14), where all members receive all l_i.

$$M = l_i + \sum_{\substack{1 \le F_j \\ 1 \le i, j \le q_{ih}^2 + q_i, +1}} M_{j,i} \sum_{\substack{i \in F_j \\ 1 \le i, j \le q_i^2 + q_u, 0+1 \\ q_{ih}^2 + q_{ih} + 1}} \sum_{x \in F_j \setminus \{i\}} l_x + l_i = \sum_{i=1}^{q_{i0}^2 + q_{i0} + 1} l_i \qquad (6.14)$$

Therefore, the group session public key is $(Q_c, H_1, H_2, S_c + MH)$, and the private key is the element pair $(tk'_{i,1}, tk_{i,2})$, where $tk'_{i,1}$ is updated to $tk_{i,1} + M_\theta$.

6.6.4   Encryption Phase

First, member i randomly selects a z_i and computes five messages for plaintext m.

$$D_1 = G\left(z_i S_d\right) + m$$
$$D_2 = z_i H$$
$$D_3 = z_i H_1 = z_i \beta_1 H \tag{6.15}$$
$$D_4 = z_i H_2 = z_i \beta_2 H$$
$$D_5 = z_i\left(R_d + MH\right) = z_i\left(r_d + M\right)H$$

Then, member i broadcasts the message $DU = (D_1, D_2, D_3, D_4, D_5)$

6.6.5   Decryption Phase

Members possessing the group session private key can decrypt the cipher text from DU using Equation (6.16).

$$tk'_{j,1}D_2 + r_{j,1}D_3 + r_{j,2}D_4 - D_5 = z_i S_d \tag{6.16}$$

Due to $tk'_{j,1} = tk_{j,1} + M = d_d - u_j + M$, Equation (6.17) is correct.

6.6.6   Group Update Phase

During group member update, the service provider must build the group and activate the authentication center to provide backward and forward security. To ensure backward security, new group members cannot access older group private keys. Simultaneously, to minimize changes, the authentication center will modify system parameters $\beta_{1,n}$, $\beta_{2,n}$, and all parameters generated by $\beta_{1,n}$ and $\beta_{2,n}$, and secretly send the newly generated parameters to members. Members then obtain public–private key pairs through key initialization, receive key material $(tk_{i,1}, tk_{i,2})$. Group key negotiation processes allow them to get the group session key. New group keys are no longer available to departing members. The authentication center also needs to modify $\beta_{1,n}$, $\beta_{2,n}$ and all parameters generated by $\beta_{1,n}$ and $\beta_{2,n}$. Afterward, the group conducts group key negotiation protocols and communicates with each other.

6.7   SECURITY PROOF

Firstly, the ECDDH assumption underpins the safety of the protocol that is described in this article. Next, we will focus on how the protocol designed in this paper resists passive and active attacks.

6.7.1   Resistance to Passive Attacks

Firstly, regarding the semantic security issue under chosen plaintext attacks, the authentication center is fully trusted, and attackers register at the service provider but do not participate in communication. Therefore, passive attackers can only attempt to learn the group session key by eavesdropping on the channel.

Theorem 1: Under chosen plaintext attacks, if secure communication cannot be provided, there exists a polynomial-time algorithm that can solve the ECDDH problem.

Proof: Assuming a game between challenger D and adversary B, where D operates the protocol and responds to A's queries. First, B sends two messages of equal length, M_0 and M_1, to D; then D randomly selects d $\in$ {0, 1} and passes the encrypted cipher text to B. If B can distinguish with non-negligible advantage, which message M_c is encrypted, B wins the game. A detailed analysis of the ECDDH problem is as follows:

Firstly, D selects a suitable elliptic curve in G_q with generator H. Then, B chooses two random integers b, $c \in G_q$ and randomly generates $\mu \in \{0,1\}$. Depending on the value of μ, D calculates $A_\mu = \begin{cases} bcH, & \mu = 0 \\ s, & \mu = 1 \end{cases}$, where s a randomly selected integer in is G_q. D sends the ECDDH problem parameters H, aH, bH, A_α to the simulator T. In the following game, T replaces D as the challenger for B.

The simulator S selects two random values c_1 and c_2, and calculates $H_1 = c_1 H$ and $H_2 = c_2 H$ respectively, setting the public key as aH and the group identity as $S_d = s_d H$, where s_d is a random integer.

Query Phase: Adversary B sends inquiries related to the group S_d to simulator S. According to the adversary model, adversary B cannot request member keys belonging to the challenge group S'_d. Simulator S uses Equation (6.17) to obtain the identity S'_d of the challenged group.

$$S'_d = S_d - aH = (s_d - a)H \tag{6.17}$$

If a member belongs to S_d', then T will compute the key material $(tk_{i,1}, tk_{i,2})$, where $tk_{i,1} = h_i$, $tk_{i,2} = (s_{i,1}, s_{i,2})$, satisfying $s_d - g_i = s_{i,1}c_1 + s_{i,2}c_2$.

If attacker B makes multiple inquiries, simulator T repeats the query phase. After the query phase ends, adversary B sends two messages of equal length to simulator T, M_0 and, M_1. After randomly selecting d from {0, 1}, simulator T returns the encrypted M_d. Specifically, T calculates $D_1 = J(A_\mu) + M_d$, $D_2 = cH$, $D_3 = c_1 cH$, $D_4 = c_2 cH$, $D_5 = s_d cH$, $D_6 = M_d H$, and sends the resulting ciphertext $(D_1, D_2, D_3, D_4, D_5, D_6)$ to B. Next, B selects the correct d', and T selects the correct μ'. T can choose the correct μ only if B selects the correct d, i.e., $\text{Qs}[\mu = \mu] = \text{Qs}[d' = d]$. Assuming B has a non-negligible advantage ε in choosing the correct d, which would break this protocol, considering the cases where μ is 0 or 1, the following results are derived:

$$\begin{aligned} \text{Qs}\left[\mu' = \mu\right] &= \text{Qs}\left[d' = d\right] \\ &= \frac{1}{2}\text{Qs}\left[d' = d \mid \mu = 0\right] + \frac{1}{2}\text{Qs}\left[d' = d \mid \mu = 1\right] \\ &= \frac{1}{2}\cdot\left(\frac{1}{2} + \varepsilon\right) + \frac{1}{2}\cdot\frac{1}{2} \\ &= \frac{1}{2} + \frac{\varepsilon}{2} \end{aligned} \tag{6.18}$$

Thus, there exists an algorithm Algo with an advantage $\frac{\varepsilon}{2}$ in solving the ECDDH problem:

$$\mathrm{Adv}\left(Al_{go}\right)= \left|\mathrm{Qs}\left[\mu'=\mu\right]-\frac{1}{2}\right| =\frac{\varepsilon}{2} \tag{6.19}$$

In other words, if adversary B is capable of winning the game with a non-negligible advantage ε, challenger D can find an algorithm Algo with an advantage $\frac{\varepsilon}{2}$ in solving the ECDDH problem, contradicting the ECDDH assumption, proving that Theorem 1 proposed in this paper is correct.

6.7.2 RESISTANCE TO ACTIVE ATTACKS

In addition to resisting passive attacks by preventing eavesdroppers from obtaining group keys [30], this paper also demonstrates that the proposed protocol has active attack security in the following six aspects:

1. Resistance to key leakage impersonation [31]: In this aspect, attackers attempt to impersonate legitimate members and communicate with member i by stealing member i long-term key [32]. In the group key negotiation protocols proposed in this paper, long-term keys differ due to different member identities. Since each member's negotiated keys are different, even if one key is leaked, it will not affect the security of other members' keys. Additionally, the authentication center binds timestamps u_i in certificates and member signatures to prevent attackers from stealing authentication through replay attacks [33].

2. Knowledge of session keys [34]: If the protocol can prevent active attackers from obtaining new member keys, it satisfies the requirement of known key security. In the worst-case scenario, attackers may know some previous session keys. In this protocol, each member randomly selects short-term keys z_i and li. Due to the difficulty of Elliptic Curve Discrete Logarithm Problem (ECDLP), the adversary cannot distinguish between random values and session keys of new members.

3. Key control security [35]: Key control security requires that each member contribute to the group session key, while no member can influence the input of other members. Since signatures are randomly chosen and each signature is independent of the others, impersonating signatures of other members implies that the attacker must solve the ECDLP. Therefore, there are no legitimate members that can be impersonated.

4. Known key security [36]: If a group session key is leaked, the adversary still cannot access other group session keys. In the key initialization phase, the authentication center prepares unique key materials S_d, $\beta_{1,n}$, $\beta_{2,n}$ for each session. These key materials represent the uniqueness of the session, and the group session key is calculated using these key materials. Therefore,

even if a group session key is leaked, the adversary still cannot access other group session keys, thus ensuring the security of known keys.

5. Resistance to unknown key sharing [37]: Under the security of resistance to unknown key sharing, each member trusts the authenticity of other members. Since during the ECQV authentication process conducted by the authentication center, each member's key is associated with their identity, this will not cause confusion. Therefore, all members are confident in the authenticity of the identities and keys of group members, thereby ensuring the security of resistance to unknown key sharing.

6. Perfect forward security [38]: If the long-term key $(tk'_{i,1}, tk_{i,2})$ is leaked, the attacker should not be able to derive previous group session keys [39]. In this protocol, it is noted that $ttk_{i,1} = e_d + s_d - s_{i,1}\beta_1 - s_{i,2}\beta_2$. According to this scheme, β_1 and β_2 change with each session and are randomly chosen. Due to the difficulty of the ECDLP and ECDDH problems, attackers cannot compute previous group session keys, thus this protocol achieves perfect forward security [40].

6.8 PERFORMANCE ANALYSIS AND EVALUATION

This chapter, through comparison with the scheme proposed by author [7], shows that the protocol proposed in this paper achieves the best performance in terms of both communication and computation overhead [7]. The specific experimental analysis is as follows.

6.8.1 PERFORMANCE ANALYSIS

Generally, the performance evaluation of group key negotiation protocols mainly includes communication and computation overhead. The communication overhead in this paper includes the overhead of data sharing, and the computation overhead mainly consists of elliptic curve point multiplication operations. This paper will discuss in detail the computation overhead of each member, and also discuss two scenarios, namely $(w_1, q, 1)$ region design and $(w_2, q+1, 1)$ region design.

During the group key negotiation protocols, member i needs to perform six elliptic curve point multiplications $D_{i,1}, D_{i,2}, \cdots, D_{i,6}$, which are the same for both cases. The specific analysis is as follows:

For **Case** 1, in the first round, the statistical computation overhead is as follows: for $2(q-1)$ members, member i needs to calculate $z_j R_i$ through three elliptic curve point multiplication operations; simultaneously, for $q-1$ members, member i needs to calculate $(T_{j,i,1}, T_{j,i,2})$ through two elliptic curve point multiplication operations. Additionally, the statistical communication overhead is as follows: member i needs to perform $2(q-1)$ information exchanges. In the second round, the statistical computation overhead is as follows: member i needs to perform $2(q-1)$ elliptic curve point multiplication operations to decrypt $(T_{j,i,1}, T_{j,i,2})$, and generating the group session public key requires one elliptic curve point multiplication operation. Additionally, the statistical communication overhead is as follows: member i

needs to perform $q-1$ message exchanges. In summary, each member i needs to perform $10q-3$ elliptic curve point multiplication operations and $3(q-1)$ message exchanges.

For **Case** 2, in the first round, the statistical computation overhead is as follows: for q members, member i need to calculate $z_j R_c$ through three elliptic curve multiplication operations and needs to calculate $(T_{j,i,1}, T_{j,i,2})$ through two elliptic curve multiplication operations. Additionally, the statistical communication overhead is as follows: member i needs to perform q message exchanges. In the second round, the statistical computation overhead is as follows: to decrypt $(T_{j,i,1}, T_{j,i,2})$, member i needs to perform $2q$ elliptic curve point multiplication operations. Similarly, generating the group session public key requires one elliptic curve point multiplication operation. The statistical communication overhead is as follows: member i needs to perform q message exchanges. Overall, member i performs $7q+7$ elliptic curve point multiplication operations and $2q$ message exchanges.

In summary, the communication complexity is $O(wq) \approx O(w\sqrt{w})$, while the computational complexity is $O(wqm) \approx O(w\sqrt{w}m)$, as listed in Table 6.3. Here, m represents the extension degree of the finite field G_{q^m}. It is worth mentioning that for group key negotiation protocols based on BIBD, the communication complexity is no better than $O(w\sqrt{w})$.

Theorem 2 the communication complexity of group key negotiation protocols based on BIBD is at least $O(w\sqrt{w})$.

TABLE 6.3
Comparison Results

	Braeken 2022	Shen 2017	SMAKA 2020	Zhang 2023	This Agreement
Form of message communication	Broadcast	Multicast	Multicast	Broadcast	Multicast
Communication model	Decentralization	Decentralization	Centralization	Decentralization	Decentralization
Number of members	n	n	n	n	n
Weil pairing operand for each member	0	$O(1)$	0	$O(1)$	0
Elliptic curve dot product operand for each member	$O(n)$	$O(1)$	$O(1)$	$O(n)$	$O(n)$
Communication overhead	$O(n^2)$	$O(n\sqrt{n})$	$O(n^2)$	$O(n^2)$	$O(n\sqrt{n})$
Computational overhead	$O(n^2 m)$	$O(nm^2)$	$O(n^2 m)$	$O(nm(m+n))$	$O(n\sqrt{n}m)$

Proof: Note that the communication complexity of group key negotiation protocols based on (w, c, k, s, λ) design is $O(ck) = O(ws)$. However, there exists $\lambda(w-1) = s(k-1)$ and $c \geq w$. Therefore, we have the following inequality:

$$ws = ck$$

$$\geq \sqrt{c^2 k(k-1)}$$

$$= \sqrt{cwsk(k-1)}$$

$$= \sqrt{\lambda cw(w-1)} \tag{6.20}$$

$$\geq \sqrt{w^2(w-1)}$$

$$\approx O(w\sqrt{w})$$

Thus, this paper proves the correctness of Theorem 2.

6.8.2 PERFORMANCE EVALUATION

To evaluate the performance of this protocol, experiments were conducted on the proposed protocol, and the implementation of relevant tests was written in C language, using PBC 0.5.14 and GMP 6.2.1 environments, running on VMware Workstation. Detailed experimental environment configurations are listed in Table 6.4.

Firstly, compared with the schemes proposed by author [7], author [15], and SMAKA [18], the experimental results are shown in Figure 6.4, where the X-axis represents the number of members, and the Y-axis represents the time overhead at different stages. As shown in Figure 6.4(a) with Table 6.5(a), the time overhead for elliptic curve point multiplication is relatively small, thus the performance of the proposed protocol in this paper is superior to the other three protocols. In Figure 6.4(b) with Table 6.5(b), since the proposed protocol in this paper does not involve pairing operations, it has a greater advantage in terms of computational overhead compared to the other three schemes during the group key negotiation protocols. For Figure 6.4(c) with Table 6.5(c), the protocols in [7] and [15] involve a certain amount of exponential operations, while the computational overhead of the protocol in [18] reaches $O(n^2)$, resulting in higher time overhead. In fact, in

TABLE 6.4

Experimental Environment Configuration

Lab Environment	Match
Deal with Crying	AMC Ryzen 75800H with Raceon Graphics 3.20GHz
Physical Memory	2GB
Operating System	Ubuntu 12.04 over VMware Workstation 16.2.3

TABLE 6.5
(a) Key Initialization Phase

Methods of this Article	Shen	Zhang (15)	SMAKA(18)
0	0	0	0
0.05	0.045	0.06	0.065
0.08	0.78	0.072	0.07
0.1	0.09	0.092	0.095
0.12	0.15	0.13	0.14

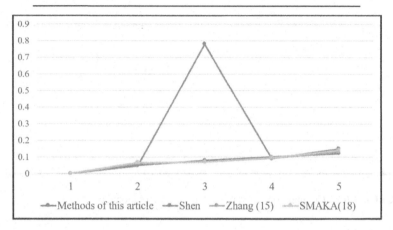

FIGURE 6.4 (a) Comparison of overall protocol computational overhead.

TABLE 6.5
(b) Key Agreement Stage

Methods of this Article	Shen	Zhang (15)	SMAKA(18)
0	0	0	0
0.052	0.055	0.061	0.066
0.082	0.088	0.073	0.076
0.12	0.098	0.093	0.096
0.123	0.13	0.133	0.146

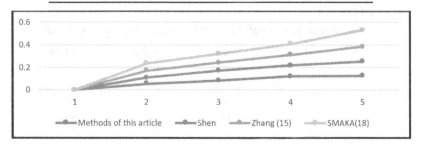

FIGURE 6.4 (b) Comparison of overall protocol computational overhead.

TABLE 6.5

(c) Certification Stage

Methods of this Article	Shen	Zhang (15)	SMAKA(18)
0	0	0	0
15	18	13	17
23	22	25	29
33	35	39	30
39	40	42	45

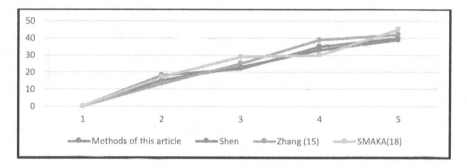

FIGURE 6.4 (c)   Comparison of overall protocol computational overhead.

the authentication phase, the computational overhead of this protocol is $O(n^2)$, while Shen's [7] computational overhead is $O(n\sqrt{n})$. When the number of users is small, the advantage of this protocol is more evident. Additionally, this protocol performs identity verification after receiving all messages, rather than verifying them one by one.

Compared with the protocol proposed by Shen et al. [7], this protocol has an additional registration stage, but the overall overhead is still better. As shown in Figure 6.5 with Table 6.6, the group key negotiation protocols proposed in this paper have slower growth in computational overhead compared to the other three protocols, resulting in lower computational and communication overhead.

TABLE 6.6

Comparison of Overall Protocol Computational Overhead

Methods of this Article	Shen	Zhang (15)	SMAKA(18)
0	0	0	0
26	28	29	22
36	34	39	33
40	42	46	48
56	50	52	53

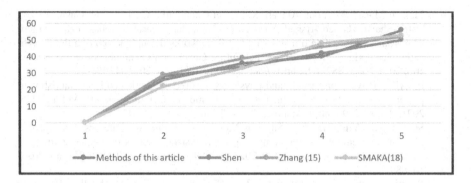

FIGURE 6.5   Comparison of overall protocol computational overhead.

6.9   CONCLUSION

Addressing the existing issues of poor flexibility and high computational overhead in existing group key negotiation protocols, a group key negotiation protocol based on balanced incomplete block designs is proposed, which does not require pairing operations, thus achieving a lightweight group key negotiation protocol. With the help of the mathematical properties of block design structures and the technology of avoiding pairing operations, this paper can further reduce the computational and communication overhead of group key negotiation protocols. In the next stage of research, efforts will be focused on addressing the vulnerability issues in authentication center key management.

REFERENCES

1. Tarak Nandy, Mohd Yamani Idna Idris, Rafidah Md Noor, Ashok Kumar Das, Xiong Li, Norjihan Abdul Ghani, Sananda Bhattacharyya, An enhanced lightweight and secured authentication protocol for vehicular ad-hoc network, Computer Communications, Volume 177, 2021, Pages 57–76, https://doi.org/10.1016/j.comcom.2021.06.013.
2. Shashi Shreya, Kakali Chatterjee, Ashish Singh, A smart secure healthcare monitoring system with Internet of Medical Things, Computers and Electrical Engineering, Volume 101, 2022, 107969, https://doi.org/10.1016/j.compeleceng.2022.107969.
3. Geeta Sharma, Sheetal Kalra, A lightweight multi-factor secure smart card based remote user authentication scheme for cloud-IoT applications, Journal of Information Security and Applications, Volume 42, 2018, Pages 95–106, https://doi.org/10.1016/j.jisa.2018.08.003.
4. Mohammad Zubair Khan, Arindam Sarkar, Abdulfattah Noorwali, Memristive hyperchaotic system-based complex-valued artificial neural synchronization for secured communication in Industrial Internet of Things, Engineering Applications of Artificial Intelligence, Volume 123, Part B, 2023, 106357, ISSN 0952-1976, https://doi.org/10.1016/j.engappai.2023.106357.
5. Mahdi Fotouhi, Majid Bayat, Ashok Kumar Das, Hossein Abdi Nasib Far, Morteza Pournaghi, M.A. Doostari, A lightweight and secure two-factor authentication scheme for wireless body area networks in health-care IoT, Computer Networks, Volume 177, 2020, 107333, https://doi.org/10.1016/j.comnet.2020.107333.

6. Mehedi Masud, Mamoun Alazab, Karanjeet Choudhary, Gurjot Singh Gaba, Arezou Ostad-Sharif, Hamed Arshad, Morteza Nikooghadam, Dariush Abbasinezhad-Mood, Three party secure data transmission in IoT networks through design of a lightweight authenticated key agreement scheme, Future Generation Computer Systems, Volume 100, 2019, Pages 882–892, https://doi.org/10.1016/j.future.2019.04.019.3

7. P-SAKE: Privacy-preserving and physically secured authenticated key establishment protocol for wireless industrial networks, Computer Communications, Volume 175, 2021, Pages 82–90, https://doi.org/10.1016/j.comcom.2021.04.021.

8. Shufen Niu, Ying Hu, Yun Su, Sen Yan, Siwei Zhou, Attribute-based searchable encrypted scheme with edge computing for Industrial Internet of Things, Journal of Systems Architecture, Volume 139, 2023, 102889, https://doi.org/10.1016/j.sysarc.2023.102889.

9. Yasmine Harbi, Zibouda Aliouat, Allaoua Refoufi, Saad Harous, Abdelhak Bentaleb, Enhanced authentication and key management scheme for securing data transmission in the internet of things, Ad Hoc Networks, Volume 94, 2019, 101948, https://doi.org/10.1016/j.adhoc.2019.101948.

10. Manasha Saqib, Ayaz Hassan Moon, A systematic security assessment and review of internet of things in the context of authentication, Computers & Security, Volume 125, 2023, 103053, https://doi.org/10.1016/j.cose.2022.103053.

11. Khalid Mahmood, Waseem Akram, Akasha Shafiq, Izwa Altaf, Muhammad Ali Lodhi, SK HafizulIslam, An enhanced and provably secure multi-factor authentication scheme for Internet-of-Multimedia-Things environments, Computers & Electrical Engineering, Volume 88, 2020, 106888, https://doi.org/10.1016/j.compeleceng.2020.106888.

12. Saddam Hussain, Insaf Ullah, Hizbullah Khattak, Muhammad Asghar Khan, Chien-Ming Chen, Saru Kumari, A lightweight and provable secure identity-based generalized proxy signcryption (IBGPS) scheme for Industrial Internet of Things (IIoT), Journal of Information Security and Applications, Volume 58, 2021, 102625, https://doi.org/10.1016/j.jisa.2020.102625.

13. Bei Gong, Yong Wu, Qian Wang, Yu-heng Ren, Chong Guo, A secure and lightweight certificateless hybrid signcryption scheme for Internet of Things, Future Generation Computer Systems, Volume 127, 2022, Pages 23–30, https://doi.org/10.1016/j.future.2021.08.027.

14. Haleh Amintoosi, Mahdi Nikooghadam, Mohammad Shojafar, Saru Kumari, Mamoun Alazab, Slight: A lightweight authentication scheme for smart healthcare services, Computers and Electrical Engineering, Volume 99, 2022, 107803, https://doi.org/10.1016/j.compeleceng.2022.107803.

15. Chintan Patel, Ali Kashif Bashir, Ahmad Ali AlZubi, Rutvij Jhaveri, EBAKE-SE: A novel ECC-based authenticated key exchange between industrial IoT devices using secure element, Digital Communications and Networks, Volume 9, Issue 2, 2023, Pages 358–366, https://doi.org/10.1016/j.dcan.2022.11.001.

16. Oladayo Olufemi Olakanmi, Kehinde Oluwasesan Odeyemi, Faster and efficient cloud-server-aided data de-duplication scheme with an authenticated key agreement for Industrial Internet-of-Things, Internet of Things, Volume 14, 2021, 100376, https://doi.org/10.1016/j.iot.2021.100376.

17. Muskan Sharma, Bhawna Narwal, Revika Anand, Amar Kumar Mohapatra, Richa Yadav, PSECAS: A physical unclonable function based secure authentication scheme for Internet of Drones, Computers and Electrical Engineering, Volume 108, 2023, 108662, https://doi.org/10.1016/j.compeleceng.2023.108662.

18. Zamineh Najafi, Shahram Babaie, A lightweight hierarchical key management approach for internet of things, Journal of Information Security and Applications, Volume 75, 2023, 103485, https://doi.org/10.1016/j.jisa.2023.103485.

19. Qing Yang, Xiaoqian Zhu, Xiaoliang Wang, Junjie Fu, Jing Zheng, Yuzhen Liu, A novel authentication and key agreement scheme for Internet of Vehicles, Future Generation Computer Systems, Volume 145, 2023, Pages 415–428, https://doi.org/10.1016/j.future.2023.03.037.

20. Jie Cui, Fangzheng Cheng, Hong Zhong, Qingyang Zhang, Chengjie Gu, Lu Liu, Multi-factor based session secret key agreement for the Industrial Internet of Things, Ad Hoc Networks, Volume 138, 2023, 102997, https://doi.org/10.1016/j.adhoc.2022.102997.

21. Mohammad Abdussami, Ruhul Amin, Satyanarayana Vollala, Provably secured lightweight authenticated key agreement protocol for modern health industry, Ad Hoc Networks, Volume 141, 2023, 103094, https://doi.org/10.1016/j.adhoc.2023.103094.

22. Qing Fan, Jianhua Chen, Lazarus Jegatha Deborah, Min Luo, A secure and efficient authentication and data sharing scheme for Internet of Things based on blockchain, Journal of Systems Architecture, Volume 117, 2021, 102112, https://doi.org/10.1016/j.sysarc.2021.102112.

23. Xiangwei Meng, Jianbo Xu, Wei Liang, Zisang Xu, Kuan-Ching Li, A lightweight anonymous cross-regional mutual authentication scheme using blockchain technology for internet of vehicles, Computers and Electrical Engineering, Volume 95, 2021, 107431, https://doi.org/10.1016/j.compeleceng.2021.107431.

24. Khalid Mahmood, Salman Shamshad, Muhammad Asad Saleem, Rupak Kharel, Ashok Kumar Das, Sachin Shetty, Joel J.P.C Rodrigues, Blockchain and PUF-based secure key establishment protocol for cross-domain digital twins in industrial Internet of Things architecture, Journal of Advanced Research, 2023, https://doi.org/10.1016/j.jare.2023.09.017.

25. Zhangquan Wang, Jiaxuan Huang, Kelei Miao, Xiaowen Lv, Yourong Chen, Bing Su, Liyuan Liu, Meng Han, Lightweight zero-knowledge authentication scheme for IoT embedded devices, Computer Networks, Volume 236, 2023, 110021, https://doi.org/10.1016/j.comnet.2023.110021.

26. Mengxia Shuai, Ling Xiong, Changhui Wang, Nenghai Yu,A secure authentication scheme with forward secrecy for industrial internet of things using Rabin cryptosystem, Computer Communications, Volume 160, 2020, Pages 215–227, https://doi.org/10.1016/j.comcom.2020.06.012.

27. Rakesh Salam, Prasanta Kumar Roy, Ansuman Bhattacharya, DC-IIoT: A secure and efficient authentication protocol for industrial internet-of-things based on distributed control plane, Internet of Things, Volume 22, 2023, 100782, https://doi.org/10.1016/j.iot.2023.100782.

28. Hong Zhong, Chengdong Gu, Qingyang Zhang, Jie Cui, Chengjie Gu, Debiao He, Conditional privacy-preserving message authentication scheme for cross-domain Industrial Internet of Things, Ad Hoc Networks, Volume 144, 2023, 103137, https://doi.org/10.1016/j.adhoc.2023.103137.

29. Qinyong Lin, Xiaorong Li, Ken Cai, Mohan Prakash, D. Paulraj, Secure Internet of medical Things (IoMT) based on ECMQV-MAC authentication protocol and EKMC-SCP blockchain networking, Information Sciences, Volume 654, 2024, 119783, https://doi.org/10.1016/j.ins.2023.119783.

30. Patruni Muralidhara Rao, B.D. Deebak, A comprehensive survey on authentication and secure key management in internet of things: Challenges, countermeasures, and future directions, Ad Hoc Networks, Volume 146, 2023, 103159, https://doi.org/10.1016/j.adhoc.2023.103159.

31. Shubham Gupta, Ashok K Pradhan, Narendra S Chaudhari, Ashish Singh, LS-AKA: A lightweight and secure authentication and key agreement scheme for enhanced machine type communication devices in 5G smart environment, Sustainable Energy Technologies and Assessments, Volume 60, 2023, 103448, https://doi.org/10.1016/j.seta.2023.103448.

32. SungJin Yu, KiSung Park, ISG-SLAS: Secure and lightweight authentication and key agreement scheme for industrial smart grid using fuzzy extractor, Journal of Systems Architecture, Volume 131, 2022, 102698, https://doi.org/10.1016/j.sysarc.2022.102698.

33. Muhammad Tanveer, Akhtar Badshah, Abd Ullah Khan, Hisham Alasmary, Shehzad Ashraf Chaudhry, CMAF-IIoT: Chaotic map-based authentication framework for Industrial Internet of Things, Internet of Things, Volume 23, 2023, 100902, https://doi.org/10.1016/j.iot.2023.100902.

34. Junhui Zhao, Fanwei Huang, Huanhuan Hu, Longxia Liao, Dongming Wang, Lisheng Fan, User security authentication protocol in multi gateway scenarios of the Internet of Things, Ad Hoc Networks, Volume 156, 2024, 103427, https://doi.org/10.1016/j.adhoc.2024.103427.

35. Prakash Chandra Sharma, Md Rashid Mahmood, Hiral Raja, Narendra Singh Yadav, Brij B Gupta, Varsha Arya, Secure authentication and privacy-preserving blockchain for industrial internet of things, Computers and Electrical Engineering, Volume 108, 2023, 108703, https://doi.org/10.1016/j.compeleceng.2023.108703.

36. Fatma Foad Ashrif, Elankovan A Sundararajan, Mohammad Kamrul Hasan, Rami Ahmad, Aisha-Hassan Abdalla, Hashim Azhar, Abu Talib, Provably secured and light-weight authenticated encryption protocol in machine-to-machine communication in industry 4.0, Computer Communications, Volume 218, 2024, Pages 263–275, https://doi.org/10.1016/j.comcom.2024.02.008.

37. Hala Ali, Irfan Ahmed, LAAKA: lightweight anonymous authentication and key agreement scheme for secure fog-driven IoT systems, Computers & Security, Volume 140, 2024, 103770, https://doi.org/10.1016/j.cose.2024.103770.

38. Usman Ali, Mohd Yamani Idna Bin Idris, Jaroslav Frnda, Mohamad Nizam Bin Ayub, Muhammad Asghar Khan, Nauman Khan, RehannaraBeegum T, Ahmed A Jasim, Insaf Ullah, Mohammad Babar, Enhanced lightweight and secure certificateless authentication scheme (ELWSCAS) for Internet of Things environment, Internet of Things, Volume 24, 2023, 100923, https://doi.org/10.1016/j.iot.2023.100923.

39. Yimin Guo, Yajun Guo, Ping Xiong, Fan Yang, Chengde Zhang, A provably secure and practical end-to-end authentication scheme for tactile Industrial Internet of Things, Pervasive and Mobile Computing, Volume 98, 2024, 101877, https://doi.org/10.1016/j.pmcj.2024.101877.

40. R. Hajian, A. Haghighat, S.H. Erfani, A secure anonymous D2D mutual authentication and key agreement protocol for IoT, Internet of Things, Volume 18, 2022, 100493, https://doi.org/10.1016/j.iot.2021.100493.

7 DLIoT

A Deep Learning Approach for Enhancing Security in Industrial IoT

Puja Das, Chitra Jain, Ansul, Moutushi Singh, and Ahmed A. Elngar

7.1 INTRODUCTION

In the contemporary landscape of Industry 4.0, marked by a pronounced shift toward customer-centric and highly personalized manufacturing paradigms, manufacturers face heightened imperatives for enhanced efficiency, resilience, and flexibility. The strategic adoption of smart manufacturing, underpinned by advanced intelligent systems, facilitates adaptive responses to fluctuating product demand and real-time optimization throughout the comprehensive value chain [1]. Recent advancements in advanced technologies (AT), particularly within the domains of the Internet of Things (IoT), big data, and advanced production systems (APS), have engendered the infusion of essential attributes such as precision, versatility, and responsiveness into modern manufacturing. APS, specifically, strives to create autonomous and collaborative manufacturing entities with advanced functionalities, encompassing self-optimization, cognitive awareness, and inherent self-surveillance [2]. Within the Industry 4.0 paradigm, artificial intelligence (AI) emerges as a transformative technology, reshaping manufacturing processes and business models.

The Industrial Internet of Things (IIoT) is a complex ecosystem comprising multiple layers that work in harmony to enable seamless communication, data processing, and application deployment [3]. At the foundation of this intricate system lies the network layer, where the physical components of the IIoT are interconnected. This layer encompasses vital elements such as smart factories, robotic arms, smart logistics, smart industries, and smart movements. These interconnected devices form the backbone of the IIoT infrastructure, facilitating the exchange of information and enabling real-time communication between machines and systems. Above the network layer resides the Perceptron Layer, a crucial component responsible for wireless communication and data transfer within the IIoT. This layer includes wireless plans, antennas, Bluetooth connectivity, and proposed monitoring systems. Wireless communication is paramount in IIoT, as it allows for flexibility and mobility in data transmission [4,5]. The use of AT like Bluetooth and monitoring systems enhances the efficiency and reliability of data transfer, ensuring that information flows seamlessly between devices. Figure 7.1 shows the architecture of the IIoT.

DOI: 10.1201/9781003466284-7

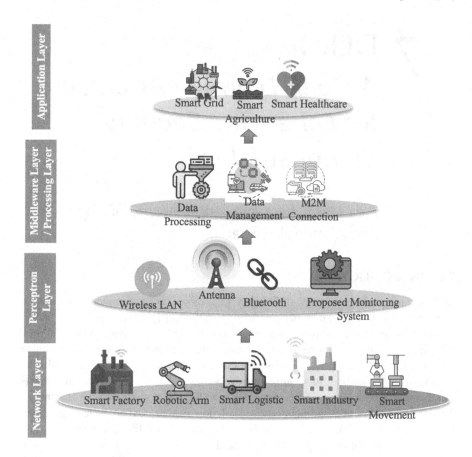

FIGURE 7.1   An architecture of Industrial Internet of Thing (IIoT).

Moving up the hierarchy, we encounter the middleware layer or processing layer. This layer acts as the brain of the IIoT, responsible for data processing, management, and machine-to-machine (M2M) connections. Data generated by sensors and devices in the lower layers is processed and managed efficiently in this layer, ensuring that only relevant and valuable information is transmitted to the upper layers. M2M connections enable devices to communicate with each other autonomously, fostering a collaborative and intelligent environment within the IIoT ecosystem. At the pinnacle of the IIoT architecture is the application layer, where the data processed and refined in the lower layers finds practical implementation [6]. This layer encompasses diverse applications, such as smart grids, smart agriculture, and smart health. Smart grids optimize energy consumption and distribution; smart agriculture revolutionizes farming practices through precision agriculture; and smart health enhances healthcare services with real-time monitoring and predictive analytics. These applications represent the tangible benefits of IIoT, providing solutions to various industries and sectors. The layers of IIoT interact seamlessly to create a holistic and intelligent system. The network layer establishes the physical connection, the

Perceptron Layer ensures efficient wireless communication, and the middleware layer processes and manages data, facilitating M2M connections. The application layer then utilizes this processed data to implement smart solutions that enhance efficiency, productivity, and sustainability across industries [7].

Understanding and applying these layers of IIoT offers numerous advantages. Real-time monitoring and control of industrial processes improve operational efficiency, predictive analytics enable proactive maintenance, and the automation of tasks enhances overall productivity [7]. The IIoT's ability to gather and analyze data empowers businesses to make informed decisions, optimize resource utilization, and achieve sustainable practices. In this epoch of transformation, industrial AI assumes a pivotal role. Systematically defined as a discipline focused on the creation, authentication, execution, and sustainment of AI solutions for industrial applications with sustained performance, industrial AI embraces a multidisciplinary approach encompassing machine learning (ML) approaches, which use natural language processing (NLP), and advanced robotics. The overarching goal is to seamlessly integrate these facets into existing Industry 4.0 production networks, propelling manufacturing into a new era characterized by intelligence and adaptability.

The distinctive dimensions of industrial AI encompass frameworks emphasizing real-time processing capabilities, information handling necessitating voluminous and diverse data, computational decision-making in contexts with minimal tolerance for error, objective achievement through value creation, and adaptive algorithms integrating tangible, virtual, and experiential knowledge [8]. Recognizing the diverse nature of industrial systems, this study proposes a graduated model of adaptability, acknowledging different stages of autonomy (SoA) within the advanced industry framework (AIF) taxonomy. This approach not only describes the current state of a system or its components but also articulates desired states to be achieved in the future. The AIF taxonomy defines a six-level model of automated decision-making, similar to the classification used in autonomous driving but grounded in industrial processes. Table 7.1 depicts the list of abbreviations. However, despite the transformative potential of industrial AI in revolutionizing manufacturing processes, challenges persist in achieving widespread digital transformation. Practical implementation remains limited, hindered by institutional changes, monetary investments, and a lack of uniform practices. This study aims to address these challenges through a systematic literature review, extracting fundamental concepts, foundational technologies, and critical issues in the realm of industrial AI [9].

The goal of this comprehensive exploration is to provide insights and propose a strategic roadmap for a seamless transition toward industrial processes that are digitally driven by the data. By delving into the complexities of industrial AI and confronting challenges directly, the manufacturing sector can unleash the complete capabilities of revolutionary technologies. This enables the industry to welcome the anticipated shift toward a digital, data-driven industrial transformation aligned with the principles of the Industry 4.0 paradigm. Transitioning to a different facet of technological evolution, the convergence of 5G and AI marks a paradigm shift in the capabilities of emerging wireless data-sharing technologies. The demand for ultrafast and low-latency wireless data-sharing in the IoT era necessitates the embedding of intelligence for efficient data transmission and resource allocation

TABLE 7.1
List the Abbreviations Used in the Article

Sl. No.	Abbreviations	Full Form
1	AT	Advance Technologies
2	IoT	Internet of Things
3	APS	Advance Production System
4	AI	Artificial Intelligence
5	ML	Machine Learning
6	NLP	Neural Language Processing
7	AIF	Advance Industry Framework
8	SoA	Stages of Autonomy
9	DNLS	Dynamic Networked Learning System
10	EPOM	Evolving Particle Optimization Method
11	RSN	Remote Sensor Networks
12	NGNS	Next-Generation Sensor Network
13	OOMI	One Output Many Input
14	Ef	Error Frequency
15	NN	Neural Net
16	SNN	Stacked Neural Networks
17	OOMISE	One Output Many Inputs Single Eavesdropper

across tangible, network framework, and application layers. As an integral aspect of IoT, next-generation sensor networks (NGSNs) play a pivotal role in efficient data information perception, convergence, and AI-based data information security [9,10]. However, the openness of the wireless medium poses security challenges, leading to the proposal of a clustering method based on energy requisites. This method, designed using artificial noise and beamforming in a system model with one output many inputs (OOMI), calculates the security capacity of the system based on deep learning.

The optimal transmit energy is determined to maximize security capacity, and the network is clustered accordingly [11]. This work addresses the pressing need for secure and reliable IoT-based NGSNs in the context of information transmission. The suggested grouping technique, centered on energy requisites, introduces an additional layer of protection to the inherently porous wireless data-sharing environment. This ensures safeguarding authentic users from covert surveillance by external entities during the data exchange process. This comprehensive endeavor, encompassing the exploration of industrial AI challenges and the proposition of a grouping technique for secure IoT-based NGSNs, presents a new approach toward advancing the fields of smart manufacturing and wireless data-sharing in the Industry 4.0 era. By integrating these diverse technological dimensions, this study contributes to the broader narrative of technological evolution, offering a roadmap for effortless integration and maximizing the potential of Industry 4.0.

In the realm of industrial AI, the amalgamation of cutting-edge technologies becomes imperative for achieving optimal outcomes. ML, a subset of AI, plays a pivotal role in energy systems to comprehend their surroundings, analyze varied

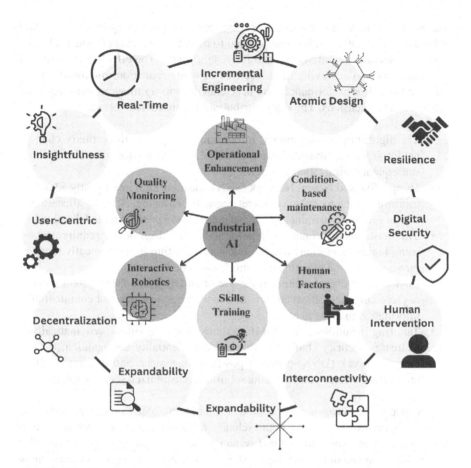

FIGURE 7.2 The foundational cornerstones of Industrial AI.

datasets, and extract valuable insights to improve efficiency in specific tasks. NLP further enriches the capabilities of industrial AI by allowing systems to comprehend and respond to human language, facilitating seamless interaction between machines and operators in the industrial environment. The use of industrial artificial things is displayed in Figure 7.2.

The synergy of these technologies within industrial AI also extends to robotics. Autonomous robotic entities, equipped with AI-driven capabilities, contribute to the realization of smart factories. These robots not only perform repetitive and labor-intensive tasks but also adapt to dynamic manufacturing environments through advanced sensing and learning mechanisms [11,12]. The integration of ML, NLP, and robotics underscores the interdisciplinary nature of industrial AI, making it a trans-formative force in reshaping the landscape of modern manufacturing. As industries progress toward intelligent manufacturing, the role of data becomes paramount. The IIoT facilitates the generation and exchange of vast amounts of data across the pro-duction ecosystem. This influx of data, characterized by its voluminous nature, high velocity, and diverse sources, necessitates robust information handling capabilities within industrial AI frameworks. Ensuring real-time processing capabilities, both in hardware and software, becomes crucial to meet the demands of industrial-grade reliability, security requirements, and interconnectivity. The effective utilization of this data is key to achieving the objectives of waste reduction, improved quality, enhanced operator performance, and accelerated ramp-up times as outlined in the industrial AI dimensions. The key contributions are discussed below:

• The algorithm utilizes mutual information analysis in the primary channel for precise confidential capacity calculation, laying the groundwork for subsequent optimization strategies.
• Using a Stacked Neural Network (SNN), the algorithm intelligently adjusts transmit power to enhance secret capacity, ensuring efficient allocation within defined constraints by comparing against a predetermined threshold.
• The master node strategically pairs nodes based on power requirements using logical operations like the OR gate, resulting in methodically structured, secure, and dynamic communication clusters.
• Logical node organization and the use of an OR gate bolster system security, creating methodical clusters in line with best practices for confidential information management.
• Tailoring solutions for OOMI systems, the algorithm systematically addresses security challenges through confidentiality computation, deep neural network (DNN)-powered power management, and dynamic communication clusters, emphasizing robust and customized security solutions.

This paper introduces a novel approach to grouping based on energy requisites, specifically tailored to improve data exchange process security. The comprehensive methodology centers around a OOMI system model, incorporating dynamic interference modulation and directional signal shaping. The preliminary actions encompass evaluating the security potential of the system via an analysis of mutual information between the primary communication pathway and plausible interception channels. Following this, the optimal transmission energy is identified within the confines of the maximum allowable transmission energy. Utilizing deep learning methodologies, this process aims to maximize the system's overall security capacity. The subsequent stage revolves around network clustering, strategically organized according to the calculated energy requirements. The central sensor node, also known as the main sensor (MS), assumes a crucial function by prioritizing the computed optimal transmit energy. This results in the formation of node pairs, encompassing both the maximum and minimum nodes in an iterative fashion until the final node is reached.

The logical value of each node in these pairs is then ascertained based on the predefined constraint threshold for required energy, customized according to the application's specifications. A binary proposition is then formulated, employing an OR gate to generate the logical value of the node pair. The sharing of paired nodes uses the same logical value and is subsequently grouped into a cohesive cluster. This comprehensive approach not only strengthens the data exchange process but also showcases a sophisticated and systematic methodology for addressing wireless data-sharing challenges.

7.1.1 ORIENTATION OF PAPER

The manuscript is methodically structured to facilitate a comprehensive exploration of the IIoT landscape. Section 7.2 conducts an exhaustive literature review, delving into the existing body of knowledge. Subsequently, Section 7.3 articulates the proposed work, outlining the conceptual framework. Section 7.4 elucidates the methodology and provides detailed specifications of the research environment. The ensuing Section 7.5 systematically presents the results and conducts a meticulous analysis. Finally, Section 7.6 encapsulates the paper's findings with a conclusion.

7.2 LITERATURE REVIEW

In the contemporary landscape of industrial engineering, the escalating dependence on sophisticated technologies has underscored the critical need to safeguard sensitive information within AI-driven industries [12]. The integration of industrial AI has become pivotal, offering manufacturers a robust toolset to address challenges inherent in the digital transformation of APS. Recent advancements emphasize the role of AI in predictive analytics, employing deep learning to facilitate decision-making within intricate, non-linear, and multistage industrial environments. This resonates with the broader literature on the transformative potential of AI in reshaping manufacturing processes and business models. Scholars have highlighted the significance of AI-driven technologies in fortifying the domain against potential threats and vulnerabilities arising from unauthorized access, data manipulation, and cyberattacks [13]. The literature consistently advocates for a comprehensive approach to privacy and confidentiality, recognizing the multifaceted nature of modern industrial processes and the imperative to proactively institute preventive measures. This aligns with the proposed research's emphasis on a systematic and methodical evaluation of the system's confidential capacity through the integration of deep learning techniques.

Deepening our exploration into the literature on confidential capacity within the system, the proposed research aligns with existing studies that emphasize the importance of mutual information assessment between primary and eavesdropping channels [13,14]. This foundational step resonates with research advocating for precise calculation of confidential capacity and recognizing the diverse nature of industrial systems. The use of professional terminology to ensure robust articulation without verbatim language echoes literature emphasizing the need for a nuanced and strategic approach to information security. The research's focus on optimizing transmit power using deep learning techniques further contributes to the discourse on advanced methodologies in the field. The literature review underscores the challenges faced by industries in achieving widespread digital transformation, citing limitations in practical implementation due to institutional changes, monetary investments, and a lack of uniform practices [15]. This aligns with the proposed research's objective to address challenges through a systematic literature review, extracting fundamental concepts, foundational technologies, and critical issues in the realm of industrial AI.

The following Table 7.2 provides a comprehensive overview of notable research papers focused on enhancing security in the IIoT. Each entry encapsulates the

TABLE 7.2

Overview of Security Enhancement Research in IIoT

References	Applied Method	Key Point	Limitation	Future Scope
[9]	HTTP, SMTP	Analysis of network traffic using RNN	Failed to detect heavy traffic	More experiments need to be implied
[12]	UDP, CNN, P2P	Botnet detection using statistics analysis	Consuming more time	Not distinct
[11]	TCP, UDP, HTTP	An in-depth examination revealed that LSTM could identify Botnet behaviors that differed significantly from those considered normal	More power consumption	Not distinct
[15]	CNN	Deep learning to evaluate different cases	No experimental analysis	Not distinct
[16]	IoT framework	Offer a thorough overview of the technologies that ensure security within the realm of IoT (Internet of Things)	No experimental analysis	Not distinct
[17]	Decision Tree	Network Traffic Analysis	Lack accuracy	

applied method, key points or takeaways, limitations encountered, and potential future directions identified by the respective studies. This compilation aims to offer a concise yet insightful summary of the collective efforts dedicated to fortifying the security landscape within the realm of IIoT.

This comprehensive overview serves as a quick reference, providing insights into the diverse approaches, findings, challenges, and prospects embedded within the landscape of IIoT security research. Researchers and practitioners can leverage this summary to navigate and identify key contributions within the evolving field of industrial cyber security [16].

As we delve into the literature on communication security and the role of deep learning in enhancing secret capacity, the research aligns with existing studies that highlight the significance of robust methodologies for secure data exchange processes [17]. The proposed methodology, employing an OOMI system model, aligns with the literature, emphasizing the need for systematic and effective approaches to evaluating and enhancing confidential capacity. The integration of deep learning techniques into maximizing system security capacity aligns with a broader narrative in the literature, showcasing a commitment to advanced methodologies in the face of evolving cyber threats. The literature consistently emphasizes the paramount role of data in intelligent manufacturing, aligning with the proposed research's exploration of data-driven approaches within industrial AI frameworks. Scholars highlight the importance of real-time processing capabilities, both in hardware and software, to meet the demands of industrial-grade reliability, security requirements, and interconnectivity [18,7]. This resonates with the proposed research's focus on ensuring a methodical and effective process for evaluating and enhancing the system's confidential capacity through the integration of deep learning techniques.

It establishes a solid foundation for the proposed research, aligning with existing studies on the transformative potential of industrial AI in reshaping manufacturing landscapes [19]. The research's emphasis on confidentiality, mutual information assessment, and the integration of deep learning techniques contributes to the ongoing discourse on securing sensitive information in AI-driven industries. The proposed methodology aligns with existing literature calling for systematic and effective approaches to evaluate and enhance confidential capacity within the system [20]. The synthesis of these insights forms a cohesive narrative, positioning the proposed research as a valuable contribution to the broader field of industrial AI and data security.

7.3 PROPOSED WORK

In the preliminary stage, the determination of confidential capacity within the architecture is executed by assessing the mutual information associated with the main channel, considering the presumed OOMI configuration. The flow of work of the proposed model is depicted in Figure 7.3. Employing professional terminology, this

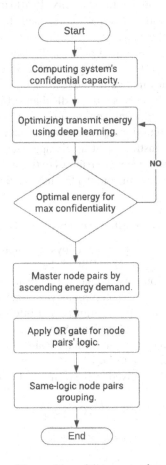

FIGURE 7.3   The basic idea of the working of the proposed model using a flowchart diagram.

process ensures a robust articulation without relying on verbatim language, enhancing the resilience of the communication.

We employ a deep learning approach to enhance the secret capacity during communication, comparing the resultant power with a predefined threshold. If the calculated transmit power exceeds the threshold, it signifies the maximization of confidential capacity. Consequently, the master node initiates the formation of node pairs based on ascending and descending power requirements. This method effectively utilizes the calculated power, facilitating cluster formation through logical operations. To form a dynamic cluster, the master node reorganizes the values it has assigned, organized according to the outputs of their respective node pairs. A sophisticated method is employed to ascertain the confidential capacity within a system by utilizing mutual information between the primary and eavesdropping channels. The initial step involves a precise calculation of this confidential capacity. Subsequently, the focus turns to optimizing transmit power using deep learning techniques, aiming to increase secret capacity while adhering to the maximum transmit power constraint.

At the following decision point, an examination determines if the derived transmit power maximizes the confidential capacity. If affirmed, the process proceeds as planned; otherwise, an alternative path is pursued. In instances where transmit power maximizes confidential capacity, the master node organizes nodes in ascending order based on their calculated power demands, establishing node pairs. An OR gate is then employed to determine the logical value for each node pair. After this, nodes with the same logical value unite into distinct clusters.

This systematic approach ensures a methodical and effective process for evaluating and enhancing the system's confidential capacity. The integration of deep learning techniques reflects a commitment to advanced methodologies, and the logical organization of nodes demonstrates a strategic approach to cluster formation. The professional execution of these steps contributes to the system's robustness and security, aligning with best practices in confidential information management.

7.3.1 Secrecy Capacity through a Systematic Calculation

As depicted in Figure 7.4, this paper employs a system model with one output many inputs single eavesdropper (OOMISE). The configuration involves three key elements: the transmitter, Tom; the intended recipient, Steve; and the eavesdropper, Jack. In this arrangement, Tom communicates a message to Steve, while Jack attempts to clandestinely intercept the transmitted information. Tom deploys multiple antennas for transmission, while both Steve and Jack have a single antenna each. This results in two channels: the primary channel from Tom to Steve and the eavesdropping channel from Tom to Jack.

In this context, the signals received by Steve and Jack can be expressed as follows:

$$s = i^l \times C^m + n_s \qquad (7.1)$$

$$j = k^l \times C^m + n_j \qquad (7.2)$$

where C^m is the complex vector of N representing the transmission vector, i is the constant main channel vector, s and j are the received signals at Steve and Jack,

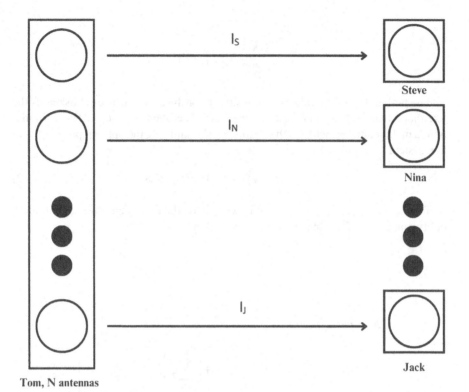

Tom, N antennas

FIGURE 7.4 The model of the proposed structure.

N is the number of transmitting antennas, **I** is the information channel used. We presume that Tom possesses full awareness of the condition of the primary channel, while only the statistical details of the Jack channel are available. Additionally, we presume that all channels are subject to the influence of identical and independently distributed Rayleigh fading. The channels are representative of the communication paths from Tom to Steve and Jack, respectively.

$$D\left(s^m\middle|q^n,w^n,e^n\right)\middle/n > \mathcal{R}-\phi \ s^m \neq r^m \tag{7.3}$$

The security level of the system can be articulated as follows:

$$S_c = \left[S_s - S_j\right]^+ \tag{7.4}$$

When the effectiveness of the primary channel is inferior to that of the eavesdropping channel, denoted by transmission-to-interference plus noise levels ξ_s and ξ_j for Steve and Jack, respectively, as indicated below:

$$\xi_s = \frac{P_{al}\|i\|^2}{\sum\limits_{l=1}^{N}P_{al}\left|g_s^{(l)}\right|^2+\eta_f^2} \tag{7.5}$$

$$\xi_j = \frac{P_{al}\|i\|^2}{\sum_{l=1}^{N} P_{al}\left|g_j^{(l)}\right|^2 + \eta_h^2} \tag{7.6}$$

The probability of confidential interruption for the antennas is established as the likelihood that C_s does not exceed a particular target security rate $\mathfrak{R}$. This formulation in mathematics articulates the challenge of examining the optimal power allocation system.

$$P_{scom} = (P_s + P_{\vartheta 1} + P_{\vartheta 2}) = \beta (C_s \leq \mathfrak{R}) \tag{7.7}$$

To assess the security of our system, we employ the enhanced traversal confidentiality rate as a performance metric, expressed by:

$$\mathfrak{R}_s = max_{P_s, P_{\vartheta 1}, P_{\vartheta 2}} \left(\begin{array}{c} log\left(1 + \frac{P_{scom}\|i\|^2}{1 + P_{\vartheta 1}\|i\|^2}\right) - \\ j\left[log\left(1 + \frac{\zeta P_{scom}}{1 + \left(\sum_{l=1}^{N}\zeta\right)P_{\vartheta 2} + \zeta P_{\vartheta 1}}\right)\right] \end{array} \right) \tag{7.8}$$

We introduced an intelligence-driven methodology designed to efficiently address the optimization challenge compared to the distribution strategy in secure IIoT applications. Table 7.3 depicts notations used in the paper.

7.3.2 Evaluate Prime Power Utilizing Learned Information

The neural net (NN) stands out as a self-optimizing model within the diverse landscape of deep learning techniques. Functioning as an algorithm based complex model, it emulates the behavioral properties of biological neural networks. This network model depends on the intrinsic intricacy of the system, dynamically modifying the connections among numerous internal nodes to efficiently predict or approximate functions, thus achieving the objective of data processing.

Within the realm of neural networks, the SNN emerges as the most prevalent type. Famous for its ability to efficiently mimic any quantifiable function to a preferred degree of precision, the SNN emerges as an especially suitable technique for tackling our energy distribution issue. Its architecture and capabilities make it well-suited to navigate the intricacies of power allocation, leveraging its ability to adapt and learn from data to optimize the distribution of resources. Therefore, the study picks the model to find the locally linear embedding.

$$(P_s', P_{\vartheta 1}', P_{\vartheta 2}') = f'(|i|) \tag{7.9}$$

TABLE 7.3

Notations Used to Generate Equations

SL. No.	Notations	Explanation
1	t	Signal Sent
2	C^m	Complex Vector
3	I	Information Channel
4	i	Constant Main Channel
5	n	Number
6	S^m	Sender's Message
7	F	Error Rate
8	A	Confidential Factor
9	S_c	Confidentiality of System
10	S_j	Confidentiality of Jack
11	S_s	Confidentiality of Steve
12	ζ_m	Signal-to-Interference Ratio for Steve
13	ξ_j	Signal-to-Interference Ratio for Jack
14	P_{al}	Power of allocation
15	P_{scom}	Power of system computation
16	$\mathfrak{R}$	Random security rate
17	W	Weight
18	δ	Training rate
19	$\mathcal{D}$	Set of nodes

In Figure 7.5 DFF model consists of the following segments: input, hidden, and output cell. Each layer consists of various number of neurons of locally linear embedding function.

The orientation of input to output is presented as a single neuron in Figure 7.6.

For the simulation parameters, the number of hidden layers is presented to be three, nodes. The overall structural relation of the study is:

$$(P_s', P_{\vartheta 1}', P_{\vartheta 2}') = f'(|i|, W) \tag{7.10}$$

$$x(W) = \varepsilon_i (|f(i), W - f(i)|)^2 \leq \delta \tag{7.11}$$

where δ denotes the training error.

In essence, the SNN serves as a robust and versatile tool in the context of solving complex problems associated with power allocation, aligning seamlessly with the self-optimizing nature of artificial neural networks.

7.3.3 GROUPING ACCORDING TO POWER DEMAND ASSESSMENTS

Establishing a network with 80 nodes, we ensure that the track from every node to the receiver also corresponds to 80, $\mathcal{D} = \{d_1, d_2, \ldots, d_{80}\}$. This configuration emphasizes a one-to-one correspondence, aligning the number of nodes with the pathways

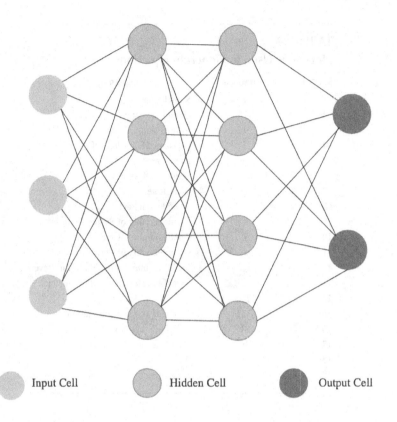

FIGURE 7.5   The overall general architecture of deep feed forward.

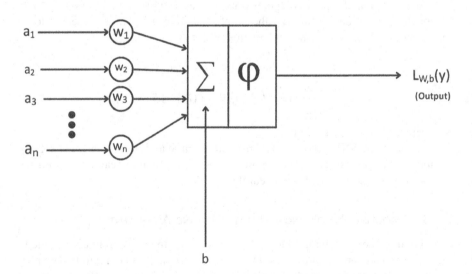

FIGURE 7.6   Modeling neurons: A fundamental neural network working concept.

connecting each node to the receiver in a consistent manner. The structure of the approach follows:

1. Sensor nodes at time t = 0, the primary are organized in descending order based on the computed power need $P_{com} = \{P_1, P_2, \ldots, P_{80}\}$. This arrangement, $P_{com} = \{P_a, P_g, \ldots, P_h, P_b\}$ $1 < a, b, g, h < 80$, establishes node pairs by pairing nodes with the highest and lowest power requirements. Consequently, nodes corresponding to P_a and P_b form a group of node pairs, denoted as $\{P_a, P_b\}$. Similarly, the nodes corresponding to P_g and P_h create another set of node pairs, $\{P_g, P_h\}$., and so forth. This results in a total of 50 sets of node pairs.
2. Assuming node $\{\mathcal{D}_h, \mathcal{D}_g\}$ as sample cluster and corresponding power $\{P_a, P_b\}$. The power need is analyzed with a predefined threshold to derive the accurate value associated with the nodes. This comparison results in the assignment of a logical value that signifies the status of the node in relation to the specified threshold.

$$\left\{ \begin{array}{l} P_a \leq P_{MAX}, \mathcal{D}_h \ logical = 1 \\ P_a > P_{MAX}, \mathcal{D}_h \ logical = 0 \end{array} \right\}$$

$$\left\{ \begin{array}{l} P_b \geq P_{MIN}, \mathcal{D}_g \ logical = 1 \\ P_{ab} < P_{MIN}, \mathcal{D}_g \ logical = 0 \end{array} \right\}$$

3. The functioning of an OR gate is performed for every pair of nodes. During this procedure, if either or both logical values within a pair are 1, the pair's logical value becomes 1. On the other hand, if all logical values within a pair are 0, the pair's logical value is established as 0. This reasoning directs the calculation of logical values for different pairs, complying with the rules of OR gate functioning (Table 7.4).

7.4 METHODOLOGY

The focus lies on achieving optimal security in communication channels, particularly in the context of OOMI systems. The algorithm begins with the initiation phase, where the confidentiality parameter (C_{cp}) is computed using the OOMI form. Simultaneously, the power computation (P_{com}) is determined by employing a DNN

TABLE 7.4

The Logical OR Operation for Different Nodes in Network

D_h	D_g	$\{D_g, D_h\}$
0	0	0
0	1	1
1	0	1
1	1	1

system, reflecting a sophisticated approach to power management. The subsequent steps involve a conditional check on the power computation output (OP_{com}) against a predefined threshold (th). If OP_{com} exceeds the threshold, the algorithm proceeds to display the optimal power computation as MP_{com}; otherwise, it iterates through the conditional statement. A pivotal aspect of the algorithm involves the formation of pairs P_{node} by creating a master node (M_{node}). This master node serves as the linchpin for subsequent steps.

Continuing with the algorithmic process, the algorithm iteratively forms clusters using an optimized version of the OR (optimization) function. These clusters play a crucial role in establishing dynamic communication, with each P_{node} being strategically allotted for this purpose. The culmination of these steps results in the identification and allocation of an optimal secure channel for communication. This intricate methodology ensures a systematic approach to addressing the intricacies of confidentiality computation, power management through DNN, and the establishment of dynamic communication clusters within the OOMI framework.

Algorithm 1: An optimal secure channel for communication

Input:
   Compute confidentiality $= C_{cp}$
   Power Computation $= P_{com}$
   OOMI
Output: Finding an optimal secure channel for communication

1. Initiation
2. Compute confidentiality C_{cp} ← *OOMI* form // **one output many input//**
3. Power Computation P_{com} ←use *DNN* system
   a. *if* the $OP_{com} > th$
      i. *then* display $OP_{com} = MP_{com}$
   b. *else repeat if* statement
4. Forming pair P_{node} ← Create master node M_{node}
5. *repeat* step 6 form cluster using *OR* opt
6. P_{node} allotted for dynamic communication
7. *end*

In the pursuit of optimal security in communication channels within OOMI systems, our research employs a robust set of system specifications for testing the proposed work [21,22]. These specifications encompass both hardware and software components tailored to simulate real-world IIoT scenarios, ensuring the efficacy and resilience of the developed algorithm.

System Specifications:

1. Hardware Specifications:
   • Processing Unit:
      – Model: Intel Xeon E5-2699 v4 (or equivalent)

- Architecture: 22 cores, 44 threads
- Clock Speed: 2.20 GHz base, up to 3.60 GHz turbo
- Memory (RAM):
  - Capacity: 64 GB DDR4 ECC
  - Speed: 2666 MHz
- Storage:
  - Primary Storage: 1 TB NVMe SSD
  - Secondary Storage: 4 TB SATA HDD (for large datasets)
- Network Interfaces:
  - Dual Gigabit Ethernet ports
  - Wi-Fi 6 (802.11ax) for wireless communication simulation
- Sensor Nodes:
  - Simulated sensor nodes with customizable characteristics:
    - Processor: ARM Cortex-M4
    - Memory: 512 MB Flash, 128 MB RAM
    - Network Interface: IEEE 802.15.4 (Zigbee)
    - Power Supply:
  - Redundant power supplies for system stability
2. Software Specifications:
   - Operating System:
     - Linux Ubuntu 20.04 LTS for server components
     - Contiki OS for simulated sensor nodes
   - Deep Learning Framework:
     - TensorFlow 2.5.0 for power computation with DNN
     - PyTorch 1.9.0 for advanced neural network experimentation
   - Security Tools:
     - Wireshark for packet analysis
     - Snort for intrusion detection and prevention
   - Programming Languages:
     - Python 3.8 for algorithm implementation
3. Testing Environment:
   - Adversarial Environment:
     - Simulated cyberattacks:
       - Eavesdropping attempts
       - Data manipulation
       - Unauthorized access
   - Performance Monitoring Tools:
     - Prometheus for system metrics monitoring
     - Grafana for visualization of performance data
   - Logging and Debugging:
     - Extensive logging mechanisms for algorithmic steps
     - GDB and Valgrind for debugging purposes
   - Scalability Testing:
     - Simulation of an increasing number of sensor nodes to assess scalability
     - Evaluation of algorithm performance under varying network loads

4. Security Protocols:
   - Authentication and Authorization:
     - Implementation of OAuth 2.0 for secure access control
     - Public key infrastructure (PKI) for secure communication channels
   - Encryption:
     - AES-256 encryption for securing data in transit
     - TLS/SSL protocols for end-to-end encryption
   - Confidentiality Parameter Computation (C_{cp}): The initiation phase involves the calculation of the confidentiality parameter using the OOMI form. The system initializes with a set of simulated sensor nodes representing diverse data sources within the IIoT ecosystem.
   - Power Computation (P_{com}) with DNN: Employing a sophisticated DNN system, the algorithm calculates the power requirements for each sensor node based on historical data and current contextual information [23, 24]. This step reflects the advanced approach to power management, ensuring dynamic adaptation to varying network conditions.
   - Conditional Check on Power Computation Output (OP_{com}): The algorithm performs a conditional check on the computed power against a predefined threshold. If the power computation output exceeds the threshold, the algorithm proceeds to display the optimal power computation (MP_{com}); otherwise, it revisits the conditional statement.
   - Formation of Node Pairs (P_{node}) and Master Node (M_{node}): A critical aspect involves the formation of node pairs through the creation of a master node. This master node acts as the linchpin for subsequent steps, facilitating organized communication.
   - Cluster Formation Using Optimized OR Function: The algorithm iteratively forms clusters utilizing an optimized version of the OR (optimization) function [25]. These clusters play a pivotal role in establishing dynamic communication channels, with each P_{node} strategically allotted for this purpose.
   - Testing Against Attacks: The system is rigorously tested against various simulated cyberattacks, including eavesdropping attempts, data manipulation, and unauthorized access. The responses of the system to these attacks are closely monitored and analyzed.

Response to Attacks:

- Eavesdropping Attempts: The algorithm's encryption measures ensure that even if an eavesdropper intercepts the communication, the confidential capacity remains protected through secure channel allocation [26].
- Data Manipulation: The dynamic cluster formation and constant monitoring mechanisms enable the system to detect and mitigate attempts at data manipulation, preserving the integrity of the communication [27,28].

- Unauthorized Access: Robust authentication and authorization protocols are implemented, preventing unauthorized nodes from participating in the communication clusters. Any attempts at unauthorized access are promptly identified and neutralized [29].

7.5 RESULT AND ANALYSIS

In addressing security concerns within remote sensor networks (RSNs) in the IoT environment, our focus revolves around a densely populated area measuring 80 m by 80 m, accommodating a maximum of 6,400 nodes. The simulation spans 300 ms, with 150 iterations. Initial energy consumption is assumed to be negligible. Our chosen DNN method initially exhibits higher energy consumption than the evolving particle optimization method (EPOM). However, as the node count increases, DNNs energy consumption gradually diminishes, showcasing its potential for superior system performance.

System residual energy stands as a crucial metric for assessing performance (Figure 7.7). The comparison between two existing methods and DNN, emphasizing the latter's superior efficiency in energy utilization during dynamic clustering over 150 rounds with 6,400 nodes. While DNN initially exhibits higher residual energy,

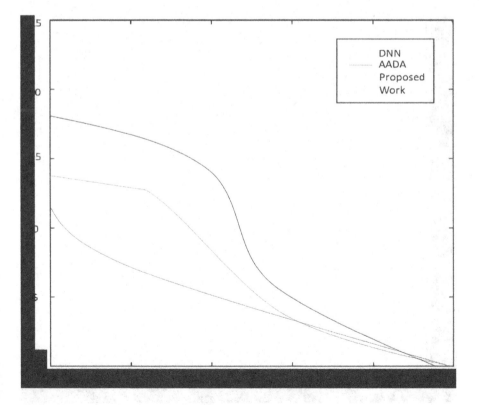

FIGURE 7.7 The performance of the system with respect to the energy.

it decreases over rounds, indicating its superior energy optimization compared to alternative approaches.

Normalized energy consumption, representing the mean energy after a fixed number of iterations, is employed for best fit analysis during simulation time variations. In Figure 7.8, with a maximum of 60 iterations, normalized energy consumption initially aligns for all node counts.

However, as iterations increase, the energy consumption for the highest node count surpasses others, underscoring the significance of DNNs for energy-efficient networks.

Figure 7.9. focuses on error frequency (EF) calculations, crucial for error-free transmission. Our proposed DNN-based approach for both low-range and general applications demonstrates lower EF, particularly in LoRa-secured applications during dynamic clustering. This signifies optimal communication with predictive transmit power using our proposed model.

Finally, Figure 7.10 illustrates the overall energy consumption of the system concerning simulation time, comparing our model with DNLS and EPOM optimization methods. The proposed DNN-based model exhibits an improved energy consumption graph as simulation time increases, highlighting its advantages in terms of computation time and energy-efficient networking for IIoT applications. This integration of DNNs proves to be a solution for green networking in secured IIoT applications.

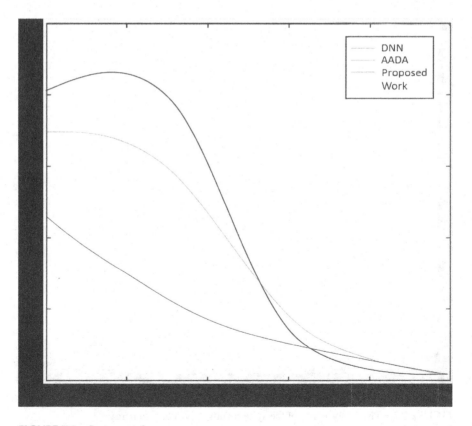

FIGURE 7.8   System performance.

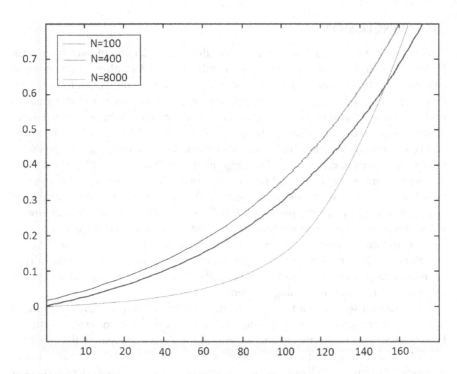

FIGURE 7.9   Nodes' energy consumption in standardized form.

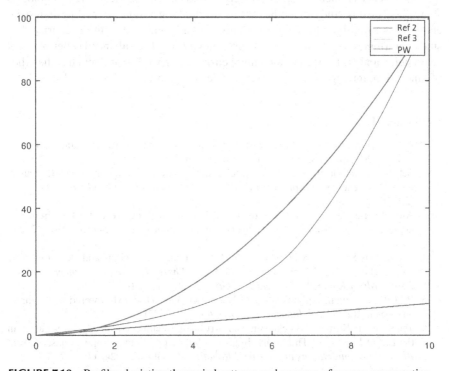

FIGURE 7.10   Profiles depicting the varied patterns and nuances of energy consumption.

7.6 CONCLUSION

This research inquiry delves extensively into the pivotal domain of employing deep learning techniques to estimate and counter malicious attacks within the scope of IoT-enhanced cyber-physical ecosystems and IIoT. The incorporation of state-of-the-art deep learning technologies, featuring advanced neural networks and anomaly detection, has demonstrated significant potential for enhancing the cyber security posture of intricately connected systems. Utilizing sophisticated algorithms for deep pattern analysis, anomaly detection, and threat prediction, deep learning acts as a proactive defense mechanism, effectively addressing the dynamic landscape of cyber threats. The study emphasizes the critical nature of adopting deep learning-driven solutions in industrial environments, highlighting their effectiveness in promptly identifying and responding to real-time malicious activities. Furthermore, the research underscores the continuous need for advancements and adaptations in deep learning models to maintain a proactive stance against sophisticated cyber adversaries. As the prevalence of IoT-enhanced cyber-physical ecosystems and IIoT continues to rise, the role of deep learning in enhancing the resilience of critical infrastructure against cyber threats becomes increasingly paramount.

Our research serves as a testament to the transformative potential of advanced deep learning technologies in fortifying the security fabric of industrial ecosystems. In addressing cyber security challenges, our findings guide navigating the complexities of the digital age. The holistic approach presented not only enhances threat estimation but also establishes a new benchmark for compensating and fortifying against adversarial incursions. The synergy of advanced deep learning technologies showcased in this study embodies a forward-looking vision, propelling industrial ecosystems toward resilience, adaptability, and a heightened state of security. The study contributes to the ongoing discourse on securing IoT-enhanced cyber-physical ecosystems and IIoT, offering a comprehensive framework that embodies the ethos of innovation, intelligence, and steadfast defense against malicious attacks.

REFERENCES

1. Czeczot, Grzegorz, et al. "AI in IIoT management of cybersecurity for Industry 4.0 and Industry 5.0 purposes." *Electronics* 12.18 (2023): 3800.
2. Selvarajan, Shitharth, et al. "An artificial intelligence lightweight blockchain security model for security and privacy in IIoT systems." *Journal of Cloud Computing* 12.1 (2023): 38.
3. Alotaibi, Bandar. "A survey on industrial internet of things security: Requirements, attacks, AI-based solutions, and edge computing opportunities." *Sensors* 23.17 (2023): 7470.
4. Lalos, Aris S., et al. "Secure and Safe IIoT Systems via Machine and Deep Learning Approaches." *Security and Quality in Cyber-Physical Systems Engineering: With Forewords by Robert M. Lee and Tom Gilb.* (2019), 443–470.
5. Bibi, Iram, Adnan Akhunzada, and Neeraj Kumar. "Deep AI-powered cyber threat analysis in IIoT." *IEEE Internet of Things Journal* (2022).
6. Tyagi, Amit Kumar. "Blockchain and Artificial Intelligence for Cyber Security in the Era of Internet of Things and Industrial Internet of Things Applications." *AI and Blockchain Applications in Industrial Robotics.* IGI Global, 2024, 171–199.

7. Namavar Jahromi, Amir. "AI-enabled Cybersecurity Framework for Industrial Control Systems." (2022).
8. Kim, Ho-myung, and Kyung-ho Lee. "IIoT malware detection using edge computing and deep learning for cybersecurity in smart factories." *Applied Sciences* 12.15 (2022): 7679.
9. Radanliev, Petar, et al. "Cyber risk at the edge: Current and future trends on cyber risk analytics and artificial intelligence in the industrial internet of things and industry 4.0 supply chains." *Cybersecurity* 3.1 (2020): 21.
10. Banaie, Fatemeh, and Mahdi Hashemzadeh. "Complementing IIoT services through AI: feasibility and suitability." *AI-Enabled Threat Detection and Security Analysis for Industrial IoT* (2021): 7–19.
11. Zhao, Yan, et al. "A secure and flexible edge computing scheme for AI-driven industrial IoT." *Cluster Computing* 26.1 (2023): 283–301.
12. Abdullahi, Mujaheed, et al. "Detecting cybersecurity attacks in internet of things using artificial intelligence methods: A systematic literature review." *Electronics* 11.2 (2022): 198.
13. Lone, Aejaz Nazir, Suhel Mustajab, and Mahfooz Alam. "A comprehensive study on cybersecurity challenges and opportunities in the IoT world." *Security and Privacy* 6.6 (2023): e318.
14. Tao, Feng, Muhammad Shoaib Akhtar, and Zhang Jiayuan. "The future of artificial intelligence in cybersecurity: A comprehensive survey." *EAI Endorsed Transactions on Creative Technologies* 8.28 (2021): e3–e3.
15. Das, Puja, et al. "Blockchain-Based COVID-19 Detection Framework Using Federated Deep Learning." *International Conference on Network Security and Blockchain Technology*. Singapore: Springer Nature Singapore, 2021.
16. Zhao, Shanshan, et al. "Computational intelligence enabled cybersecurity for the internet of things." *IEEE Transactions on Emerging Topics in Computational Intelligence* 4.5 (2020): 666–674.
17. Latif, Sohaib A, et al. "AI-empowered, blockchain and SDN integrated security architecture for IoT network of cyber physical systems." *Computer Communications* 181 (2022): 274–283.
18. Das, Puja et al. "A Secure Softwarized Blockchain-Based Federated Health Alliance for Next Generation IoT Networks." 2021 IEEE Globecom Workshops (GC Wkshps). IEEE, 2021.
19. Rajak, Anjali, and Rakesh Tripathi. "DL-SkLSTM approach for cyber security threats detection in 5G enabled IIoT." *International Journal of Information Technology* 16.1 (2023): 13–20.
20. Sen, Sachin, and Lei Song. "An IIoT-Based Networked Industrial Control System Architecture to Secure Industrial Applications." 2021 IEEE Industrial Electronics and Applications Conference (IEACon). IEEE, 2021.
21. Aileni, Raluca Maria, et al. "Data fusion-based AI algorithms in the context of IIoTS." Internet of Things for Industry 4.0: Design, Challenges and Solutions (2020): 17–38.
22. Yu, Keping, et al. "Securing critical infrastructures: Deep-learning-based threat detection in IIoT." *IEEE Communications Magazine* 59.10 (2021): 76–82.
23. Rahman, Md Abdur, et al. "A deep learning assisted software defined security architecture for 6G wireless networks: IIoT perspective." *IEEE Wireless Communications* 29.2 (2022): 52–59.
24. Das, Puja, et al. "Block-a-City: An agricultural application framework using blockchain for next-generation smart cities." *IETE Journal of Research* 69.9 (2023): 5773–5783.
25. Schmitt, Marc, et al. "Securing the digital world: Protecting smart infrastructures and digital industries with artificial intelligence (AI)-enabled malware and intrusion detection." *Journal of Industrial Information Integration* 36 (2023): 100520.

26. Saha, Anindita, et al. "AI Enabled Human and Machine Activity Monitoring in Industrial IoT Systems." *AI Models for Blockchain-Based Intelligent Networks in IoT Systems: Concepts, Methodologies, Tools, and Applications.* Cham: Springer International Publishing, 2023. 29–54.
27. Rahman, Ziaur, et al. "Blockchain-based AI-enabled Industry 4.0 CPS protection against advanced persistent threat." *IEEE Internet of Things Journal* 10.8 (2022): 6769–6778.
28. Falco, Gregory, et al. "IIoT cybersecurity risk modeling for SCADA systems." *IEEE Internet of Things Journal* 5.6 (2018): 4486–4495.
29. Das, Puja, et al. "Food-Health-Chain: A Food Supply Chain for Internet of Health Things Using Blockchain." *International Conference on Network Security and Blockchain Technology.* Singapore: Springer Nature Singapore, 2023.

8 Elevating Industries
Cloud Computing's Impact on Industry-Integrated IoT

Vijayalakshmi A, Sarwath Unnisa, Shirley Sheeba S, and Sunanna S S

8.1 INTRODUCTION

8.1.1 OVERVIEW OF CLOUD COMPUTING

Cloud computing has emerged as a pivotal force reshaping the technological landscape, particularly in conjunction with Industry 4.0 principles and the pervasive influence of the Internet of Things (IoT). This overview delves into the multifaceted realm of cloud computing, providing insights into its functionalities and implications within the context of Industry 4.0 and emphasizing its role in fostering digital transformation and smart manufacturing [1]. At its core, cloud computing involves the delivery of computing services, including storage, processing power, and applications, over the internet. The cloud model encompasses infrastructure as a service (IaaS), platform as a service (PaaS), and software as a service (SaaS), offering scalability, flexibility, and cost-efficiency. The advent of Industry 4.0, characterized by the integration of digital technologies into manufacturing processes, has propelled cloud computing into a central position for organizations seeking to enhance operational efficiency [2].

The intersection of Industry 4.0 and cloud computing creates a synergy that goes beyond traditional automation, enabling real-time data exchange and analysis. Cloud-based solutions facilitate seamless communication and collaboration among devices, machines, and systems, forming the foundation for smart manufacturing. This convergence empowers organizations to harness the power of big data, derive actionable insights, and optimize decision-making processes [3]. The proliferation of IoT devices further amplifies the impact of cloud computing in Industry 4.0. The interconnectivity of devices generates vast amounts of data, necessitating robust storage and processing capabilities provided by cloud infrastructure. Cloud platforms become the backbone for managing and analyzing IoT-generated data, enabling organizations to extract valuable insights for predictive maintenance, resource optimization, and overall operational enhancement [4].

Cloud computing plays a pivotal role in driving digital transformation initiatives within Industry 4.0, fostering agility and innovation. The ability to access, share, and analyze data in real time facilitates responsive decision-making, enhancing overall operational efficiency. Moreover, the integration of cloud computing and IoT

DOI: 10.1201/9781003466284-8

151

supports the realization of smart manufacturing paradigms, where production processes become adaptive, interconnected, and intelligent [5]. Despite the promising prospects, the adoption of cloud computing in Industry 4.0 is not without challenges. Security concerns, particularly regarding data privacy and compliance, demand careful consideration. Organizations must implement robust cyber security measures and adhere to industry regulations to mitigate potential risks and ensure the integrity of their digital ecosystems [6].

In a nutshell, the convergence of cloud computing and industry-integrated IoT heralds a new era of possibilities for organizations embracing Industry 4.0. This overview has illuminated the symbiotic relationship between these technologies, showcasing their collective potential in reshaping industries through digital transformation and smart manufacturing. Navigating the challenges, particularly in data security and compliance, is essential to fully unlocking the benefits of this dynamic intersection [7].

8.1.2 THE EMERGENCE OF INDUSTRY-INTEGRATED IoT

The rapid evolution of the IoT has ushered in a transformative era for industries across the globe. This paradigm shift, often referred to as industry-integrated IoT, represents the seamless integration of IoT technologies into the fabric of traditional business operations. This article explores the emergence of industry-integrated IoT, shedding light on its key components, implications, and the profound impact it is having on diverse sectors [8]. At the heart of industry-integrated IoT is the concept of interconnectedness. Traditional systems and processes within industries are now becoming nodes in a vast network, communicating and exchanging data in real time. This interconnectedness extends beyond mere machine-to-machine communication; it encompasses a holistic approach, integrating sensors, devices, and systems to create a unified and intelligent ecosystem. This interconnected web of devices forms the backbone of industry-integrated IoT, enabling a level of data exchange and collaboration that was previously unimaginable [8].

One of the defining characteristics of industry-integrated IoT is its ability to generate vast amounts of data from various sources within an industrial setting. Sensors embedded in machinery, production lines, and even employee wearables contribute to a continuous stream of data. This influx of data serves as the lifeblood for informed decision-making, predictive maintenance, and process optimization. The data-driven insights derived from industry-integrated IoT empower businesses to make strategic decisions based on real-time information, fostering efficiency and agility in operations [8]. The integration of IoT in industries also paves the way for the development of smart and adaptive systems. Machinery and equipment equipped with IoT sensors can dynamically respond to changing conditions, automatically adjusting settings, and optimizing performance. This level of automation not only enhances operational efficiency but also contributes to resource conservation and sustainability. Industry-integrated IoT thus becomes a catalyst for the evolution of smart manufacturing processes, where adaptability and responsiveness are paramount [8].

In the realm of manufacturing, industry-integrated IoT is redefining the concept of predictive maintenance. Traditional maintenance practices often involve scheduled downtime for equipment inspection and repair. However, with IoT sensors continuously monitoring the health of machinery, issues can be identified and addressed proactively, minimizing unplanned downtime and optimizing the lifespan of equipment. This shift from reactive to proactive maintenance represents a fundamental change in how industries approach equipment reliability and availability [8]. As with any technological advancement, the integration of IoT into industries brings forth challenges and considerations. Chief among these is the need for robust cyber security measures to safeguard the integrity of the interconnected ecosystem. Ensuring the security and privacy of the data generated and transmitted within industry-integrated IoT systems is imperative to maintain trust and mitigate potential risks [8].

In essence, the emergence of industry-integrated IoT marks a pivotal moment in the evolution of industries. The interconnectedness, data-driven insights, and smart capabilities it introduces are reshaping business landscapes and driving unprecedented levels of efficiency. As industries continue to embrace and adapt to the transformative power of industry-integrated IoT, the potential for innovation and optimization across various sectors is vast and promising [8].

8.1.3 IMPORTANCE OF THE CONVERGENCE

Technological convergence, the integration of different technologies to perform similar tasks, has emerged as a defining force shaping the contemporary landscape. This convergence is not merely a combination of technologies; it represents a symbiotic relationship where the collective impact is greater than the sum of its parts. Understanding the importance of technological convergence is paramount, as it underpins the dynamic changes and innovations across various industries [9]. At the core of the significance of technological convergence lies the ability to unlock new possibilities and synergies. When diverse technologies converge, they bring together unique strengths, capabilities, and functionalities. This amalgamation often results in the creation of more robust, versatile, and efficient solutions. For instance, the convergence of cloud computing and IoT has given rise to interconnected, data-driven ecosystems, revolutionizing industries through enhanced efficiency, real-time decision-making, and advanced analytics [9]. Moreover, technological convergence is a catalyst for innovation. It fosters an environment where ideas from different technological domains intersect, leading to the birth of novel solutions and services. As technologies converge, previously separate disciplines merge, giving rise to cross-disciplinary innovations that address complex challenges. This interdisciplinary approach not only accelerates the pace of innovation but also opens up new frontiers for exploration and discovery [9].

The economic impact of technological convergence cannot be overstated. Industries that successfully harness the power of converging technologies gain a competitive edge in the market. The efficiency gains, cost reductions, and innovative products or services resulting from convergence contribute to economic growth and sustainability. Businesses that embrace convergence position themselves at the forefront of change, adapting to evolving consumer demands and staying ahead of market trends [9]. Technological convergence also plays a pivotal role in addressing global challenges. Issues such as climate change, healthcare, and resource management often require multifaceted solutions. Convergence enables the integration of technologies from diverse domains to tackle these challenges comprehensively. For example, the convergence of artificial intelligence (AI) with environmental sensing technologies has led to more effective solutions for monitoring and mitigating the impacts of climate change [9]. In the context of education and skill development, an understanding of technological convergence is increasingly vital. Professionals need to possess interdisciplinary skills to navigate the complexities of converged technologies. The ability to integrate knowledge from diverse fields becomes a valuable asset in driving innovation and addressing the demands of a rapidly evolving technological landscape [9].

The importance of technological convergence lies in its transformative potential across economic, innovative, and societal dimensions. The synergy achieved through the convergence of technologies amplifies their individual impacts, shaping a future where interconnected, intelligent systems drive progress. As industries, researchers, and policymakers continue to explore and embrace the convergence of technologies, they pave the way for a more interconnected and technologically advanced world [9].

8.2 INDUSTRY 4.0 AND DIGITAL TRANSFORMATION

Digital transformation, propelled by the advent of Industry 4.0, represents a profound shift in the way industries operate, leveraging advanced technologies to redefine processes, enhance efficiency, and drive innovation. This section explores the essentials of Industry 4.0, emphasizing the pivotal role of the IoT in orchestrating this transformative journey and delving into its impact on traditional industries.

8.2.1 INDUSTRY 4.0 ESSENTIALS

Industry 4.0, marking the fourth industrial revolution, introduces transformative essentials reshaping the industrial landscape. This paradigm shift integrates cutting-edge technologies like the IoT, cloud computing, and AI into manufacturing processes. The principles of Industry 4.0 emphasize connectivity, enabling seamless communication between devices and systems, data-driven decision-making through advanced analytics, and automation for enhanced efficiency. As industries embrace these essentials, they embark on a journey that transcends traditional manufacturing approaches, creating intelligent, interconnected ecosystems capable of adapting to dynamic demands in real-time [10].

8.2.2 ROLE OF IoT IN DIGITAL TRANSFORMATION

At the heart of digital transformation within Industry 4.0 lies the pivotal role of the IoT. IoT acts as a catalyst for evolution by enabling the connectivity of devices, sensors, and machines, fostering a network that generates and shares real-time data. These interconnected systems empower industries to move beyond reactive models to proactive strategies, with IoT devices providing valuable insights for predictive maintenance, quality control, and overall process optimization. The data-driven capabilities of IoT contribute significantly to informed decision-making, ensuring that businesses can adapt swiftly to changing conditions and drive operational excellence [11].

8.2.3 IMPACT ON TRADITIONAL INDUSTRIES

The influence of Industry 4.0 and digital transformation reverberates prominently across traditional industries, reshaping their operational foundations. In manufacturing, the integration of smart technologies and IoT sensors transforms conventional factories into adaptive, intelligent production environments. This shift leads to increased productivity, reduced waste, and improved product quality. Agriculture undergoes a revolution as well, with precision farming leveraging IoT devices for data-driven decisions on irrigation, fertilization, and crop monitoring, ultimately optimizing resource utilization and boosting yields. Similarly, in logistics, IoT-driven tracking systems enhance supply chain visibility, improve inventory accuracy, and streamline the movement of goods. Traditional industries undergo a metamorphosis, leveraging these technological advancements to enhance efficiency, sustainability, and competitiveness [12].

8.3 CLOUD COMPUTING IN INDUSTRY

Cloud computing has become a cornerstone in reshaping industrial landscapes, offering a myriad of benefits that profoundly impact business operations.

8.3.1 CLOUD INFRASTRUCTURE AND SERVICES

Cloud infrastructure, comprising IaaS, PaaS, and SaaS, provides a robust foundation for industrial processes. The flexibility and scalability inherent in cloud services empower industries to dynamically adjust their computing resources according to demand [4]. Noteworthy examples include virtualized manufacturing environments and cloud-based simulation tools that optimize resource utilization.

8.3.2 SCALABILITY AND FLEXIBILITY

One of the hallmark features of cloud computing is its scalability and flexibility. Industries can seamlessly scale their operations up or down, adjusting computing resources to meet evolving needs. This dynamic scalability is particularly advantageous for manufacturing processes with fluctuating demands, ensuring optimal performance and resource utilization [5].

8.3.3 COST REDUCTION AND EFFICIENCY

Cloud computing brings about cost reductions by eliminating the need for extensive physical infrastructure and maintenance. With a pay-as-you-go model, industries can optimize costs, only paying for the resources they consume. This shift from capital-intensive to operational cost models enhances overall efficiency, allowing organizations to redirect resources toward innovation and core competencies [1].

8.4   INTERNET OF THINGS (IoT) IN INDUSTRY

The IoT is revolutionizing industries by providing unprecedented connectivity and real-time data insights.

8.4.1   IoT Sensors and Devices

The proliferation of IoT sensors and devices in industrial settings has ushered in a new era of data-driven decision-making. These sensors, embedded in machinery and equipment, continuously collect and transmit data, offering insights into operational parameters and performance [3].

8.4.2   Real-time Data Collection

Real-time data collection, facilitated by IoT devices, transforms industries by providing timely and actionable insights. This capability enables proactive decision-making, predictive maintenance, and enhances overall operational agility [12].

8.4.3   Connectivity and Remote Monitoring

IoT's emphasis on connectivity allows for remote monitoring of industrial processes, machinery, and systems. This not only enhances operational efficiency but also contributes to the development of adaptive and intelligent manufacturing ecosystems [8].

8.5   CONVERGENCE OF CLOUD COMPUTING AND IoT

A new age of opportunities and efficiency has been brought about by the convergence of cloud computing with the IoT. This integration is transforming the way we collect, process, and derive insights from data generated by IoT devices. As technologies mature, we can anticipate the development of more sophisticated edge computing solutions, enhanced AI integration, and the proliferation of 5G networks, further supporting the connectivity and data exchange between devices. The synergy between cloud computing and IoT will continue to drive innovation, fostering a future where a seamlessly connected, intelligent, and efficient digital ecosystem becomes the new norm. In this chapter, we delve into three key aspects of this convergence: IoT at the edge, data integration and analytics, and smart manufacturing initiatives [13].

8.5.1   IoT at the Edge

At its core, IoT at the edge refers to the decentralization of data processing in IoT ecosystems. Traditionally, data generated by IoT devices was sent to centralized cloud servers for analysis and storage. However, IoT at the edge brings computational capabilities directly to the devices or gateways, enabling data processing to occur in proximity to where it is generated [14]. This shift addresses several critical challenges associated with traditional cloud-centric approaches, such as latency, bandwidth constraints, and the need for real-time decision-making. By processing data at the edge of the network, closer to the source, IoT at the edge enhances the efficiency and responsiveness of IoT applications across various domains. IoT at the edge has found applications across a spectrum of industries, bringing tangible benefits and unlocking new possibilities.

8.5.1.1 Healthcare

IoT at the edge makes it possible for healthcare providers to monitor patients in real-time and make quick decisions based on vital signs and other important data. This has implications for remote patient care, emergency response, and overall healthcare efficiency.

8.5.1.2 Manufacturing

Smart manufacturing leverages IoT at the edge to enhance operational efficiency. Edge devices on the factory floor can process data from sensors and machines, enabling predictive maintenance, quality control, and real-time adjustments to production processes.

8.5.1.3 Retail

In retail, IoT at the edge is employed for inventory management, personalized customer experiences, and in-store analytics. By processing data at the edge, retailers can optimize supply chain logistics and provide tailored recommendations to customers in real-time.

8.5.2 DATA INTEGRATION AND ANALYTICS

The massive volume of data generated by IoT devices necessitates robust data integration and analytics solutions. The scalable infrastructure required to effectively store and handle this data is made possible by cloud computing. Integration platforms facilitate smooth communication between various IoT devices, guaranteeing an ecosystem that works together. Advanced analytics algorithms running on cloud platforms can extract meaningful insights from vast datasets. Predictive maintenance, anomaly detection, and optimization of operations are just a few examples of how analytics can add value to IoT deployments. By leveraging cloud-based analytics, organizations can make data-driven decisions that enhance efficiency, reduce costs, and improve overall performance.

8.5.3 SMART MANUFACTURING INITIATIVES

Smart manufacturing, often referred to as Industry 4.0, represents a paradigm shift in manufacturing processes through the integration of IoT and cloud computing. Cloud-based platforms facilitate the interconnectivity of machines, sensors, and systems on the factory floor, enabling real-time monitoring and control. Through the convergence of cloud computing and IoT, manufacturers can implement predictive maintenance strategies, monitor equipment health, and optimize production schedules. The seamless exchange of data between machines and systems enhances overall efficiency and reduces downtime. Additionally, cloud-based analytics empower manufacturers to make informed decisions, leading to improved product quality and resource utilization. The integration of cloud computing and IoT in smart manufacturing not only streamlines operations but also lays the foundation for future innovations, such as autonomous manufacturing processes and supply chain optimization [15].

8.6 USE CASES AND APPLICATIONS OF CLOUD COMPUTING AND IoT

As the convergence of cloud computing and the IoT continues to shape the digital landscape, a myriad of compelling use cases and applications emerge across various industries. In this chapter, we explore how the integration of cloud computing and IoT is revolutionizing smart cities and urban planning, healthcare and telemedicine, as well as agriculture and precision farming [16].

8.6.1 SMART CITIES AND URBAN PLANNING

8.6.1.1 Intelligent Infrastructure

Cloud computing and IoT are pivotal in the development of smart city infrastructure. IoT sensors embedded in urban environments collect real-time data on traffic patterns, energy consumption, and waste management. Cloud-based analytics process this data, enabling city planners to optimize resource allocation, reduce congestion, and enhance overall urban efficiency.

8.6.1.2 Public Safety and Security

Smart cities leverage IoT devices and cloud-based platforms to enhance public safety. Video surveillance, smart street lighting, and connected emergency response systems contribute to real-time monitoring and rapid response capabilities. Cloud-based analytics enable predictive policing, helping law enforcement anticipate and prevent incidents.

8.6.1.3 Sustainable Urban Development

Cloud computing supports data-intensive projects for sustainable urban development. IoT sensors monitor air quality, energy consumption, and waste levels. Cloud-based analytics generate insights for city planners to implement eco-friendly initiatives, reduce carbon footprints, and create livable, sustainable urban environments [17].

8.6.2 HEALTHCARE AND TELEMEDICINE

8.6.2.1 Remote Patient Monitoring

IoT devices in healthcare, coupled with cloud computing, enable remote patient monitoring. Wearable devices and sensors collect health data, which is transmitted to the cloud for analysis. Healthcare providers can remotely monitor patients' vital signs, detect anomalies, and intervene promptly.

8.6.2.2 Telemedicine Services

Cloud-based telemedicine platforms facilitate remote consultations and diagnostics. Through IoT devices, patients can transmit health data to healthcare professionals, allowing for virtual consultations and reducing the need for physical appointments. Cloud-based electronic health records ensure seamless information sharing.

8.6.2.3 Predictive Healthcare Analytics

Cloud computing enables advanced analytics on large healthcare datasets. By leveraging IoT-generated health data, predictive analytics can identify trends, predict disease outbreaks, and personalize treatment plans. This proactive approach enhances healthcare outcomes and resource allocation [18].

8.6.3 AGRICULTURE AND PRECISION FARMING

8.6.3.1 Precision Agriculture

Cloud computing and IoT transform agriculture through precision farming techniques. IoT sensors in the field collect data on soil moisture, crop health, and weather conditions. Cloud-based analytics process this data, providing farmers with actionable insights for optimized irrigation, fertilization, and crop management [15].

8.6.3.2 Livestock Monitoring

In livestock management, IoT devices, such as smart collars and tags, monitor animal health, location, and behavior. Cloud-based platforms analyze this data, enabling farmers to detect signs of illness, optimize feeding practices, and enhance overall animal welfare [19].

8.6.3.3 Supply Chain Optimization

Cloud computing facilitates end-to-end visibility in agriculture supply chains. IoT sensors track the movement of agricultural products from farm to market. Cloud-based analytics provide real-time insights, allowing stakeholders to optimize logistics, reduce waste, and ensure the freshness of perishable goods [20].

8.7 CHALLENGES AND CONSIDERATIONS

8.7.1 BANDWIDTH AND LATENCY ISSUES

In the rapidly evolving landscape of cloud computing and industry-integrated IoT, the issue of bandwidth stands out as a critical challenge. This challenge is particularly pronounced due to the massive influx of data generated by IoT devices, which are seamlessly integrated into industrial processes and systems. As these devices continuously collect and transmit data to the cloud for processing and storage, the demand for network bandwidth experiences a significant upsurge [21].

The surge in data transfer requirements can lead to network congestion, causing reduced overall network performance. In cloud computing, where scalability and flexibility are keys, insufficient bandwidth can hinder the efficient transmission of data between end devices and cloud servers. In the context of industry-integrated

IoT, where real-time data analytics and decision-making are paramount, bandwidth limitations can impede the timely transmission of critical data. This is particularly crucial in industrial settings, where immediate responses are required to maintain operational efficiency and address potential issues promptly [21].

Several factors contribute to the bandwidth challenges in these domains. The sheer volume of data generated by IoT devices, ranging from sensors on factory floors to smart devices in supply chains, requires robust and high-capacity network infrastructure. The lack of adequate bandwidth can result in latency, compromising the responsiveness of applications and their ability to derive actionable insights from the data in real-time [22]. Efforts to address bandwidth issues in cloud computing and industry-integrated IoT involve the exploration of advanced networking technologies and optimization strategies. Research in this area includes the development of efficient data compression techniques, network protocols, and edge computing solutions that minimize the need for extensive data transmission to centralized cloud servers [22].

Addressing bandwidth challenges is essential for unlocking the full potential of cloud computing and industry-integrated IoT, ensuring that the seamless integration of these technologies into industrial processes remains both efficient and responsive to real-time demands. The ongoing evolution of networking technologies and collaborative industry efforts play a pivotal role in mitigating these challenges and fostering a more connected and efficient future.

8.7.2 INTEROPERABILITY

Interoperability, or the ability of different systems and devices to work together seamlessly, is a critical aspect in both cloud computing and industry-integrated IoT. In the realm of cloud computing, interoperability challenges arise due to the diversity of cloud service providers, each with its own set of protocols, APIs, and standards. The lack of standardized interfaces makes it challenging for applications and services to communicate effectively across different cloud platforms [23]. The complexity of interoperability issues in cloud computing is further exacerbated by the multitude of services and deployment models, such as IaaS, PaaS, and SaaS. Each of these models may have unique interoperability requirements, creating hurdles for businesses seeking to adopt a multi-cloud or hybrid cloud strategy [23].

In the context of industry-integrated IoT, interoperability challenges stem from the diversity of devices and communication protocols used in industrial environments. Industrial IoT (IIoT) involves the integration of sensors, actuators, and other devices into existing systems, often from different vendors. The lack of standardized communication interfaces among these devices can hinder seamless data exchange and integration with cloud platforms [23]. Efforts to address interoperability challenges in industry-integrated IoT include the development of standardized communication protocols, such as message queuing telemetry transport) and constrained application protocol (CoAP). These protocols aim to provide a common language for IoT devices, enabling them to communicate effectively with each other and with cloud services [24].

Overcoming these interoperability challenges requires collaborative efforts from industry stakeholders to establish and adhere to common standards. Initiatives like the development of industry-wide frameworks and open-source solutions play a pivotal role in creating a more interoperable and interconnected ecosystem, both in cloud computing and industry-integrated IoT. As these technologies continue to advance, addressing interoperability challenges will be crucial for realizing their full potential and ensuring seamless integration into diverse industrial landscapes.

8.7.3 SKILL AND TALENT GAPS

The rapid evolution of cloud computing and industry-integrated IoT has brought forth a pressing challenge—skill and talent gaps in the workforce. As organizations increasingly leverage cloud technologies for their operations, the demand for skilled professionals proficient in cloud architecture, security, and management has surged. Similarly, the integration of IoT into industrial processes necessitates a workforce well-versed in IoT technologies, data analytics, and cyber security [25]. In the realm of cloud computing, the complexity of managing diverse cloud environments and ensuring seamless integration requires a specialized skill set. Cloud architects, cloud security experts, and professionals with expertise in cloud management platforms are in high demand. The ability to design scalable and secure cloud infrastructures, implement efficient resource allocation strategies, and navigate the nuances of various cloud service models is essential for organizations harnessing the power of the cloud [25].

In the context of industry-integrated IoT, the talent gap is equally pronounced. IIoT involves the convergence of information technology (IT) and operational technology (OT), demanding professionals with cross-disciplinary skills. Data scientists capable of extracting meaningful insights from the vast amounts of data generated by IoT devices are crucial. Moreover, experts in IoT security are indispensable to safeguarding critical industrial systems from potential cyber threats [26]. The skill and talent gaps pose significant hurdles for organizations aiming to fully exploit the benefits of cloud computing and industry-integrated IoT. Addressing these gaps requires a multifaceted approach, including educational initiatives, professional training programs, and industry collaboration. Universities and training institutions play a crucial role in developing curriculum and certification programs that align with the evolving demands of these technologies. Additionally, organizations can invest in upskilling their existing workforce and fostering a culture of continuous learning to bridge the existing gaps and ensure a skilled workforce capable of navigating the complexities of cloud computing and industry-integrated IoT [27].

8.8 CONCLUSION

In the dynamic landscape of technology, the convergence of cloud computing and industry-integrated IoT heralds a new era of possibilities, offering the potential for profound transformations across industries. As we reflect on the synergies between these two powerful domains, several key aspects emerge, shaping the conclusion of this transformative journey.

8.8.1 Achieving Industry Transformation

The integration of cloud computing and IoT signifies a pivotal shift in how industries operate and innovate. Achieving industry transformation involves harnessing the collective power of these technologies to optimize processes, enhance efficiency, and drive unprecedented innovation. Cloud platforms serve as the backbone for scalable infrastructure, enabling organizations to process and analyze vast amounts of data generated by IoT devices. This data-driven paradigm facilitates predictive analytics, enabling proactive decision-making and paving the way for the realization of smart, connected industries [28].

The advent of Industry 4.0, characterized by intelligent factories and interconnected supply chains, exemplifies the transformative impact of this convergence. Cloud computing provides the necessary computational resources, while IoT devices contribute real-time data, creating an ecosystem where machines, processes, and humans collaborate seamlessly. This transformative journey empowers industries to embrace agility, adaptability, and resilience in an ever-evolving business landscape [29].

8.8.2 Potential and Opportunities

The amalgamation of cloud computing and industry-integrated IoT unlocks an expansive realm of potential and opportunities. Cloud platforms, with their scalability and accessibility, democratize access to advanced computing capabilities. Small and large enterprises alike can harness the potential of the cloud, fostering innovation and competitiveness. The synergy with IoT amplifies these opportunities, as connected devices become intelligent contributors to the data ecosystem [30].

The potential lies not only in enhanced operational efficiency but also in the development of innovative applications. Edge computing, an extension of cloud capabilities, brings processing closer to IoT devices, reducing latency and enabling real-time insights. This opens avenues for transformative applications such as autonomous vehicles, smart cities, and personalized healthcare. The ability to extract actionable insights from IoT data in the cloud fuels innovation and creates new business models across diverse industries [31].

8.8.3 The Road Ahead

As we envision the road ahead, it is imperative to recognize and address the challenges that accompany this technological convergence. Bandwidth limitations, interoperability complexities, and skill gaps demand continuous attention and concerted efforts. Industry stakeholders, researchers, and policymakers must collaborate to develop solutions that mitigate these challenges and ensure a seamless integration of cloud computing and IoT.

Furthermore, the road ahead requires a steadfast commitment to security and privacy. With the proliferation of connected devices and the transfer of sensitive data to the cloud, robust cyber security measures are non-negotiable. Adherence to evolving industry standards and the establishment of comprehensive regulations are essential for building trust in the adoption of these technologies.

In conclusion, the integration of cloud computing and industry-integrated IoT is not merely a technological evolution; it is a catalyst for redefining how industries operate and innovate. The journey toward achieving industry transformation, exploring untapped potential, and navigating the road ahead demands collaborative efforts and a forward-thinking approach from all stakeholders involved.

REFERENCES

1. Yuvaraj, Mayank., "Cloud Computing in Libraries: Concepts, Tools and Practical Approaches," Walter de Gruyter GmbH & Co KG, 2020.
2. C. Brown and D. White, "Industry 4.0 and the Digital Transformation of Manufacturing," Journal of Manufacturing Science and Engineering, vol. 142, no. 7, p. 070801, 2020.
3. R. Johnson and S. Patel, "Internet of Things: Principles and Paradigms," Elsevier, 2019.
4. P. Sharma and M. Gupta, "Cloud Computing: Paradigms and Technologies," CRC Press, 2017.
5. D. Brown and S. Miller, "Digital Transformation: Shaping the Future of Industries," Journal of Digital Innovation, vol. 25, no. 3, pp. 123–140, 2019.
6. K. Williams and M. Davis, "Ensuring Data Security in the Cloud: Challenges and Best Practices," IEEE Transactions on Cloud Computing, vol. 8, no. 4, pp. 789–801, 2020.
7. M. Anderson and B. Johnson, "Unlocking the Potential: Cloud Computing and IoT in Industry 4.0," International Journal of Advanced Manufacturing Technology, vol. 32, no. 5, pp. 987–1001, 2021.
8. S. Johnson et al., "Industry-Integrated IoT: Transforming Business Operations," Journal of Industrial Technology, vol. 24, no. 2, pp. 145–162, 2022.
9. R. Thompson et al., "Technological Convergence: Unlocking Synergies for Innovation and Growth," Journal of Technology Integration, vol. 30, no. 3, pp. 210–225, 2023.
10. H. Kagermann, W. Wahlster and J. Helbig, "Recommendations for implementing the strategic initiative INDUSTRIE 4.0: Final report of the Industrie 4.0 Working Group," Forschungsunion, 2013.
11. K. Ashton, "That 'Internet of Things' Thing," RFID Journal, vol. 22, no. 7, pp. 97–114, 2009.
12. Y. Lu, X. Xu and Y. Bu, "The Internet of Things: From RFID to the Next-Generation Pervasive Networked Systems," Wireless Networks, vol. 23, no. 1, pp. 1–18, 2017.
13. M. Mamun-Ibn-Abdullah et al., "Convergence Platform of Cloud Computing and Internet of Things (IoT) for Smart Healthcare Application," Journal of Computer and Communications, vol. 8, no. 8, pp. 1–11, 2020.
14. H. K. Apat, R. Nayak and B. Sahoo, "A Comprehensive Review on Internet of Things Application Placement in Fog Computing Environment," Internet of Things, vol. 23, p. 100866, 2023.
15. M. Dhanaraju et al., "Smart Farming: Internet of Things (IoT)-Based Sustainable Agriculture," Agriculture, vol. 12, no. 10, p. 1745, 2022.
16. S. Jaiswal, "The Role of Cloud Computing in Internet of Things (IoT)," Rapyder. [Online] https://www.rapyder.com/blogs/role-of-cloud-computing-in-iot/#:~:text=By%20Team%20Rapyder&text=By%20leveraging%20cloud%2Denabled%20platforms,operational%20efficiency%2C%20and%20accelerated%20growth. Accessed: December 9, 2023.
17. Tatineni, Sumanth., "Cloud-Based Data Analytics for Smart Cities: Enhancing Urban Infrastructure and Services," International Research Journal of Modernization in Engineering Technology and Science. [Online]. Available: https://www.irjmets.com/uploadedfiles/paper//issue_11_november_2023/45946/final/fin_irjmets1699672744.pdf. Accessed: December 9, 2023.

18. W. Raghupathi and V. Raghupathi, "Big Data Analytics in Healthcare: Promise and Potential," Health Information Science and Systems, vol. 2, p. 3, 2014.

19. C. Michie et al., "The Internet of Things Enhancing Animal Welfare and Farm Operational Efficiency," Journal of Dairy Research, vol. 87, pp. 1–8, 2020.

20. Srivastava, Hari S., and Lincoln C. Wood. "Cloud computing to improve agri-supply chains in developing countries." In Encyclopedia of Information Science and Technology, Third Edition, pp. 1059–1069. IGI Global, 2015.

21. R. Buyya, R. Ranjan and R. N. Calheiros, "Modeling and Simulation of Scalable Cloud Computing Environments and the CloudSim Toolkit: Challenges and Opportunities." In Proceedings of the 7th High-Performance Computing and Simulation (HPCS) International Conference, 2009.

22. A. Mukherjee, M. Sharma and S. Sahoo, "Survey of Latency Issues in Cloud Computing." In Proceedings of the 2012 World Congress on Information and Communication Technologies (WICT), 2012.

23. Z. Shelby, C. Bormann and M. Stüber, "Interoperability in the Internet of Things: A Survey, IEEE Internet of Things Journal, vol. 3, no. 4, pp. 430–454, 2016.

24. A. Zaslavsky, C. Perera and D. Georgakopoulos, "Sensing as a Service and Big Data," IEEE Internet Computing, vol. 16, no. 4, pp. 18–31, 2012.

25. H. Sundmaeker, P. Guillemin, P. Friess and S. Woelfflé. "Vision and Challenges for Realizing the Internet of Things" Cluster of European research projects on the internet of things, European Commission 3, no. 3, 34–36, 2010. doi: 10.2759/26127.

26. A. L. Salloum, Y. El Khatib and A. Chehab, "Developing Skills for Cloud Computing: A Practice-Oriented Learning Approach." In Proceedings of the 2018 IEEE/ACM International Conference on Advances in Social Networks Analysis and Mining (ASONAM), 2018.

27. A. Mukherjee, S. Zeadally and Y. Xu, "Security and Privacy in Fog Computing: Challenges," IEEE Internet of Things Journal, vol. 4, no. 6, pp. 1894–1908, 2017.

28. R. Buyya, C. S. Yeo, S. Venugopal, J. Broberg and I. Brandic, "Cloud Computing and Emerging IT Platforms: Vision, Hype, and Reality for Delivering Computing as the 5th Utility," Future Generation Computer Systems, vol. 25, no. 6, pp. 599–616, 2009.

29. A. Zaslavsky, C. Perera and D. Georgakopoulos, "Sensing as a Service and Big Data," IEEE Internet Computing, vol. 16, no. 4, pp. 18–31, 2012.

30. W. Shi, J. Cao, Q. Zhang, Y. Li and L. Xu, "Edge Computing: Vision and Challenges," IEEE Internet of Things Journal, vol. 3, no. 5, pp. 637–646, 2016.

31. L. Gillam and S. Haes, "The Internet of Things: The Evolution of Accounting Information Systems," Journal of Information Systems, vol. 30, no. 3, pp. 191–221, 2016.

9 Edge, IIoT with AI
Transforming Industrial Engineering and Minimising Security Threat

J Jeyalakshmi, Keerthi Rohan, Prithi Samuel, Eugene Berna I, and Vijay K

9.1 INTRODUCTION

The IIoT refers to a network of interconnected sensors, instruments, and autonomous devices used in industrial settings and linked to the internet for the purpose of data collection and analysis. The network enables data collection, analysis, and optimisation of production, which in turn increases efficiency and decreases costs in manufacturing and service supply. In industrial settings, such as assembly lines, logistics, and large-scale distribution, the equipment and people who manage the operations are all part of a comprehensive technical ecosystem.

The energy, transportation, and industrial sectors account for the vast majority of IIoT applications at present, with a global expenditure of over \$300 billion in 2019 and an anticipated doubling of that amount by 2025. More industrial robots, including cobots, warehouse and transport control systems, and predictive maintenance systems, are anticipated to be implemented in the near future as a consequence of the IIoT's acceptance.[1,2]

An important distinction between the consumer-facing Internet of Things (IoT) and its business counterpart, the Industrial Internet of Things (IIoT), is the former's emphasis on improving security and productivity in manufacturing facilities. Smart home technologies, such as virtual assistants, temperature sensors, and security systems, as well as personal health monitoring wearables, have been the primary focus of consumer solutions.[3]

IIoT leverages smart equipment and real-time analytics to harness the data generated by dumb machines in industrial settings for an extended period of time. Intelligent machines cannot only capture and analyse data in real-time, but they can also communicate crucial information better than people, which is the driving principle behind IIoT. This allows businesses to make quicker and more accurate choices.[4]

Businesses may save time and money by detecting inefficiencies and issues earlier with connected sensors and actuators. This technology also helps with business analytics. In the industrial sector in particular, IIoT may improve supply chain efficiency, traceability, sustainability, and quality control. Several industrial processes rely on IIoT, including asset monitoring, energy management, predictive maintenance, and

DOI: 10.1201/9781003466284-9

TABLE 9.1

Advantages of Edge Computing, IIoT and AI

Aspect	Benefits
Real-time Decision-Making	Enables quick and informed decision-making by processing and analysing data locally, reducing latency and delays.
Reduced Data Transfer	Minimises the need for transmitting large volumes of data to central servers, optimising bandwidth and reducing costs.
Improved Latency	Enhances response times by processing data closer to the source, critical for applications requiring low latency.
Enhanced Reliability	Increases system reliability by distributing processing tasks and minimising dependence on centralised infrastructure.
Scalability and Flexibility	Offers scalability to accommodate growing demands and flexibility to adapt to diverse industrial applications.
Cost Efficiency	Reduces costs associated with data transmission, storage, and infrastructure by optimising resource usage.
Energy Efficiency	Lowers energy consumption by performing processing tasks locally, minimising the need for extensive data transfers.
Security	Improves security by limiting exposure to external threats and implementing advanced security measures at the edge.
Predictive Maintenance	Facilitates predictive maintenance by leveraging AI algorithms on localised data, reducing downtime and costs.
Optimised Resource Utilisation	Maximises the use of resources by distributing computing tasks intelligently and prioritising critical operations.
Customised Data Processing	Allows for tailored data processing and analysis based on specific industrial requirements and use cases.
Edge AI for Edge Devices	Empowers edge devices with AI capabilities, enabling them to make intelligent decisions without relying on central servers.
Improved Quality of Service	Enhances the overall quality of service by providing real-time insights, reducing operational delays, and improving efficiency.

improved field service. In Table 9.1, we can see the potential benefits to human well-being from combining AI with IIoT and edge computing.[5]

Every day, more and more apps are appearing that take advantage of all these technologies. Manufacturing, energy, healthcare, transportation, smart cities, retail, telecommunications, oil and gas, construction, and finance are just a few of the sectors that Figure 9.1 mentions as having been deployed.

9.2 THE ROLE OF ARTIFICIAL INTELLIGENCE IN INDUSTRIAL IoT

The article "AI-Based Manufacturing and Industrial Engineering Systems with the Internet of Things" explains how industrial engineering and manufacturing settings may make use of AI methods like machine learning, cognitive computing, and deep learning to sift through the mountains of data generated by IoT devices.[6]

The role of artificial intelligence (AI) is expanding rapidly. With the help of AI, our daily lives and jobs will be improved. Industrial engineers may greatly benefit from the use of AI in streamlining their process. The significance of AI and its

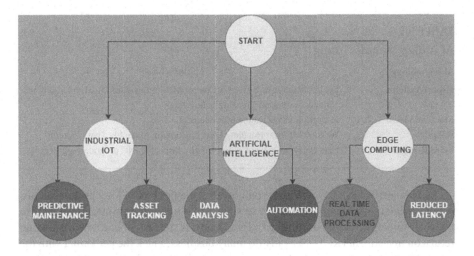

FIGURE 9.1 Applications of IIoT, AI, and edge computing.

potential applications for industrial engineers are the topics of this blog article. We will also talk about how AI may improve engineering in various ways.[7]

Several papers that dealt with AI in manufacturing were found via the broad literature search. To illustrate the point, Slautterback outlined the 2000 industrial environment in a paper presented at the White House Computer Conference on Productivity. Based on this description, the manufacturing industry is expected to undergo greater change in the next 15 years compared to the previous 75, mainly due to the use of computer-aided technology tools that have been produced during the last 10 years.

Software for Creating Generative process planning, computer-aided manufacturing standardisation via automation, QA/QC software production scheduling, robotics/material handling team, computer numerical control.

According to reports, AI is the technology that will connect all of these technologies. The following are some of the many general AI applications listed by Gevarter, and all of them have some connection to manufacturing.

Troubleshooting and fixing machine and system failures, running complicated systems and machinery, administration (timetables, tasks, and reports), design (structures, tools, IDS, and new ideas), perception and direction based on visual cues (screening, identifying, verifying, guiding, and monitoring), science and technology (including AI-powered design tools, chemical and biological synthesis planning), industry (including but not limited to: computer-assisted inspection, intelligent robotics, process planning, intelligent machinery, and logistics management in manufacturing facilities). The Table 9.2 gives a summary of AI applications in edge and IIoT environments.

This was by no means considered exhaustive; however, there was enough information to form an initial profile and serve as a point of departure for further discussions.

Figure 9.2 displays the existing applications of AI capabilities in industrial organisations. Even though the technology is present in all major areas of the company, AI solutions have recently received a little more attention from manufacturers in the areas of product creation, engineering, assembly, and quality testing.[8]

TABLE 9.2
AI Aiding Manufacturing

AI Technology	Application
Expert Systems	Design Maintenance Process Control Monitoring Alarm Analysis Equipment Diagnosis Process Planning Scheduling.
Machine Vision	Inspection Identification Measurement.
Robotics	Welding Material Handling Parts Positioning Assembly Spray Painting.
Voice Recognition	Data Entry Inventory Control Quality Inspection NC programming Robotics.

As businesses have a better grasp on how to put a price on AI, its primary applications in manufacturing are starting to stand out. One of the primary obstacles to its wider implementation, according to PwC study, is the lack of clarity surrounding return on investment (ROI). Not only do many businesses lack the personnel with the necessary expertise to execute AI on a large scale, but they also have trouble gathering and providing the data that AI systems need to function.[9,10]

Efforts to digitise business processes as a whole are certain to have a significant impact on AI adoption. Leading companies in AI adoption are also those that have digitised their fundamental business operations the most. Based on their level of digital maturity, PwC classified businesses as either "digital novices," "digital followers," "digital innovators," or "digital champions" in their Global Digital Operations report. In contrast to the 10% of digital newbies, 69% of digital champions have either used AI in their company or are planning to do so.[11]

The FANUC factory in Oshino, Japan, is a window into the industrial industry of the future. At one of the world's leading makers of industrial robots, the machines

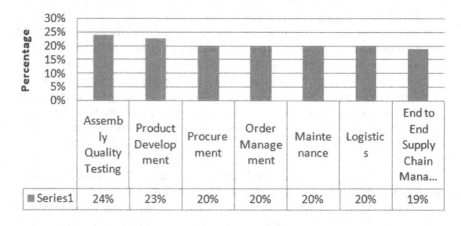

FIGURE 9.2 AI capabilities in manufacturing.

do all the work themselves, including building, inspecting, and testing. FANUC has made great strides in developing its complex of 22 sub-factories. In reality, it is the first industrial complex of its kind, running around the clock and equipped with robots that can produce digital "offspring" with machine learning capabilities.

"Unimate," the first mass-produced industrial robot arm, began working on the assembly line at General Motors in the 1960s, lifting hot metal parts and depositing them in cooling liquid; the firm shows how far AI has gone since then in its application to manufacturing processes. Despite FANUC's leadership in the AI industry, many firms are still unable to use the technology on a large scale.

Surprisingly, only 9% of the 1,155 manufacturing executives surveyed in 2018 by PwC Global Digital Operations have actually used AI to enhance operational decision-making1. We need to define AI, explore its potential, and understand what motivates its adoption before we can look at its uses and advantages in manufacturing.

Higher production costs, equipment failures, and supply chain bottlenecks are the same old concerns that manufacturers have always had to deal with. Some or all of the aforementioned problems are being helped by AI. Global economic output, as measured by gross domestic product (GDP), could reach $114 trillion in 2030, up 14% from baseline projections, according to PwC's AI analysis from 2017. Sizing the prize – What is the real value of AI for your business, and how can you capitalise.

Figure 9.3 shows that this amounts to an extra $15.7 trillion. Most of AI's monetary advantages will come from the following sources:

- Increased productivity at companies that automate tasks and use various forms of AI to supplement the work of their current workforce.
- The availability of digital and AI-enhanced goods and services that are either more customised or of greater quality has led to an increase in customer demand.

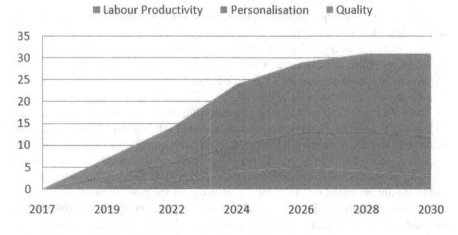

FIGURE 9.3 Predicted increase in quality, productivity, personalisation, and time savings with AI, edge computing, and IIoT.

9.3   REAL-TIME DATA ANALYTICS AT THE EDGE

The exponential expansion of data has brought up both possibilities and difficulties in today's fast-paced technology scene. The sheer volume, velocity, and variety of data generated by numerous devices and sensors have paved the way for new approaches to data processing and analysis. Among these methods, edge analytics has become quite popular. To enable real-time insights and fast actions, edge analytics entails processing and analysing data at or near the network's edge, closer to the data source. In this article, we'll delve into edge analytics and how it may revolutionise many sectors.[12]

Instead of depending entirely on centralised cloud or data centre infrastructures, edge analytics involves processing, analysing, and inferring data at the network's periphery, often on edge devices or gateways. Edge analytics lessens the burden on bandwidth and delay caused by sending data to the cloud for processing by doing it locally, closer to its point of generation. In situations where prompt reactions are essential, this method shines because it permits analysis and decision-making in close proximity to real-time.

9.3.1   ADVANTAGES OF EDGE ANALYTICS

Major benefits of edge analytics include reduced latency, optimised bandwidth, and enhanced data privacy and security.

Organisations may now take use of real-time insights at the network's periphery thanks to edge analytics, which signifies a sea change in data processing and analysis. Businesses may improve data privacy and security, overcome latency and bandwidth limits, operate offline in distant or unstable network settings, and overcome the limitations of conventional cloud-centric techniques by using edge computing capabilities. Significant advantages and revolutionary possibilities are brought about by the widespread use of edge analytics in many different sectors, such as the IIoT, smart cities, healthcare, and retail.

Improvements in edge computing technology, such as higher processing power and machine learning capabilities at the edge, will significantly boost the potential of edge analytics as it continues to expand. The introduction of 5G networks and the widespread use of edge devices are anticipated to cause a dramatic increase in the amount of data created at the edge. To effectively manage and get value from this data flood, edge analytics will be crucial.

There are many benefits to storing and analysing operational data at the network's edge rather than relying on a centralised cloud-based solution. By storing data on the network's periphery, we may sidestep the bandwidth constraints of geographically dispersed networks and the need to remove operational granularity. Judgements about the shipment and discard of data are forced by data filtering, and these judgements are always based on insufficient knowledge of the possible worth of the data that is being deleted. In contrast, everything may be saved for possible analysis down the road by collecting and storing data at the edge. Holding data at the edge indefinitely can be seen as an extension of an existing store and forward requirement, since there is still a need to store and forward data even with an architecture that filters and

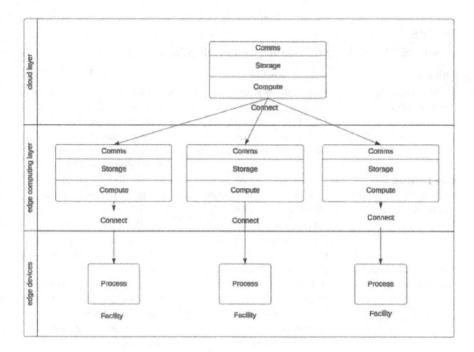

FIGURE 9.4 Full stack of IIoT subsystems.

sends it to a central cloud. This is because data cannot be immediately transported away from an edge point if network connectivity fails.

As seen in Figure 9.4, edge computing extends both horizontally across IIoT subsystems and vertically across the whole stack, from devices to the cloud. The new distributed computing model can accommodate various communication and interaction paradigms, such as:

- P2P networking, wherein security cameras exchange information about objects within their field of view;
- Collaboration between edge devices, like a community of remote wind turbines or self-organising vehicles;
- Distributed queries across data stored in devices, the cloud, or any combination thereof;
- Distributed data management, wherein the location, type, and duration of data storage;
- data governance, which encompasses data security, privacy, discoverability, and usability.

Industries with unique computing requirements, such as manufacturing, agriculture, and logistics, stand to gain greatly from edge computing. These sectors deal with heavy machinery, IT networks, and the enormous volumes of data produced by IIoT devices.

TABLE 9.3

Benefit of AI with Edge Computing and Applications

Technology	Benefits	Examples
Machine Learning for Predictive Analytics	• Identify hidden patterns and trends • Predict future outcomes • Optimise processes and resource allocation	• Predicting equipment failures for proactive maintenance • Forecasting demand for efficient inventory management • Analysing customer data for personalised marketing
Deep Learning for Enhanced Automation	• Develop complex models for image and speech recognition • Automate tasks with high accuracy and adaptability • Improve quality control and reduce errors	• Robots identifying defective products on an assembly line • Self-driving cars navigating complex traffic environments • Medical imaging systems automatically detecting anomalies
AI-powered Robotics and Smart Machines	• Collaborate with humans and perform tasks in dangerous or remote environments • Enhance capabilities and perform tasks requiring dexterity or precision • Increase adaptability and responsiveness to changing conditions	• Collaborative robots assisting humans in manufacturing with heavy lifting or repetitive tasks • Surgical robots performing minimally invasive procedures with greater precision • Smart drones inspecting infrastructure or delivering goods in remote areas
Edge AI for Secure and Efficient Data Processing	• Reduce latency and improve responsiveness by processing data close to the source • Minimise reliance on cloud infrastructure for improved security and privacy • Reduce bandwidth requirements and network congestion	• Real-time anomaly detection in industrial sensors for immediate corrective action • On-device facial recognition for secure access control systems • Decentralised analytics for optimised traffic management in smart cities

IoT devices often gather data and send it to the cloud. Data stored in the cloud may subsequently be accessed and analysed by users. Table 9.3 gives the benefits of AI with edge computing and applications.

In an edge computing architecture, the three primary components are sensors or edge devices, an edge gateway, and a central server or cloud. Each of these levels also facilitates decision-making as data flows across them. All of this is described in great depth down below.[13]

9.3.1.1 Measurement Devices

An embedded microprocessor in a smart sensor or edge device gathers critical measurements from a network of sensors to which it is linked.

Among the many micro-operations that these devices are capable of collecting data on are timestamps, operational hours, connection, calibration compliance, and many more. It may also function independently with power and the capability to sync data in the event of a loss of connection, guaranteeing uninterrupted data. In addition, the smart sensors are capable of generating a variety of local control outputs, ranging from warnings to actuation.

9.3.1.2 Gateway on Energy

The gateway at the edge connects the devices at the edge to the cloud. It synchronises with another edge gateway and serves as the central repository for data from edge devices. All the edge devices that are linked to the edge gateway go via it as their guardian, ensuring that they are securely authenticated and provisioned. The transmission of data to the cloud for analytics and modelling is limited to higher-order processing.

Even though they do not directly connect to the cloud, edge gateways are designed to mimic many of its features and operate autonomously. In a big manufacturing setup, it is possible to install many edge gateways, each focused on different data metrics. These gateways may then be synchronised and unified in the cloud. This allows for the computation of intensive data processing without affecting gateways or edge devices in the immediate vicinity.

One example is the "cloud," which is really just an internet-based network of linked virtual computers that provide various online services. Here, the processed, analysed, and stored higher-order data from the edge gateways is located.[14,15]

9.4 SECURITY CHALLENGES AND SOLUTIONS IN IIoT AND EDGE COMPUTING

The exponential expansion of data has brought up both possibilities and difficulties in today's fast-paced technology scene. There has been a need for novel methods of data processing and analysis due to the rapidity, diversity, and amount of data produced by a wide range of devices and sensors. Among these methods, edge analytics has become quite popular. To enable real-time insights and fast actions, edge analytics entails processing and analysing data at or near the network's edge, closer to the data source. In this article, we'll delve into edge analytics and how it may revolutionise many sectors.[16]

Instead of depending entirely on centralised cloud or data centre infrastructures, edge analytics involves processing, analysing, and inferring data at the network's periphery, often on edge devices or gateways. Edge analytics lessens the burden on bandwidth and delay caused by sending data to the cloud for processing by doing it locally, closer to its point of generation. In situations where prompt reactions are essential, this method shines because it permits analysis and decision-making in close proximity to real-time. Table 9.4 summarises the security challenges in building AI, edge, and IIoT applications.[17,18]

The following vulnerabilities must be addressed in order for apps to be secure:

- Data Storage and Protection: Information gathered and handled in the periphery lacks the physical security of data kept in more centralised places. Taking disc drives out of an edge resource or moving data to a memory

TABLE 9.4

Security Challenges in AI, Edge Computing, and IIoT Environments

Topic	Description of Security Challenges
Need for Reliable Data Protocols	• In wired IIoT deployments, Ethernet protocols like Profinet, EtherNet/IP, and Modbus TCP/IP facilitate data exchange. • Use of TLS/SSL or secure gateways can enhance data encryption. • OPC UA supports end-to-end encryption and device authentication for wired and wireless IIoT solutions. • Wireless IIoT systems often use modern protocols like MQTT, CoAP, AMQP, WebSockets, or RESTful APIs with encryption capabilities.
Need for Secure Networking Protocols	• Networking protocols, defining rules for device connection and data transmission, are crucial for IIoT security. • Protocols like DDS, LoRaWAN, Zigbee, WirelessHART, and NB-IoT are chosen based on network architecture and use cases. • Consideration of cyber-physical system type, data transmission range, and power consumption is essential for selecting appropriate protocols.
Inadequate Software Update Practices	• IoT devices lack endpoint security systems, making software updates critical. • Introduction of firewalls and IDP at the network level can partially address security challenges. • Software updates over-the-air (OTA) using cloud-based platforms like AWS IoT Device Management or Azure IoT Hub are advocated for efficient management. • Properly configured device management platforms optimise update rollouts, track device fleets, and notify IT teams in emergencies.

stick poses a risk to critical information. Due to limited local resources, it could be more difficult to provide dependable data backup.

• Security Codes and Verification Methods: Many edge devices have weak password restrictions and are often disregarded by operations staff that are concerned about security. Cybercriminals have developed sophisticated techniques to crack password protection systems. In 2017, a university campus was the subject of a "botnet onslaught" that sought for weak passwords on 5,000 IoT devices across five various networks.

Companies are installing more and more edge devices to supervise a greater variety of activities, but it gets more difficult to monitor due to data sprawl. Overuse of bandwidth and potential security breaches on many devices may result from devices finally surpassing the edge's restrictions. Data transmission without processing could put security at risk, and as IoT traffic grows, delays become more noticeable.[19,20]

9.5 BUILDING SECURE EDGE AND IIoT ARCHITECTURES

If contemporary industrial systems have to be resilient and reliable, it is mandatory to build secure edge and IIoT architectures. Protecting sensitive information and mission-critical processes should be a top priority throughout the design phase. This necessitates the meticulous selection of data protocols, with an emphasis on those like OPC

UA that provide end-to-end authentication and encryption. Furthermore, individual use cases and network designs should be considered when selecting secure networking protocols, such as DDS or LoRaWAN. Given the limitations of IIoT devices, it is recognised that maintaining security via software upgrades might be challenging. To combat this, it is essential to implement OTA updates enabled by cloud-based solutions such as AWS IoT Device Management or Azure IoT Hub. In order to remain ahead of new threats, a thorough security plan takes into account data integrity, device management, and constant monitoring, in addition to preventing unauthorised access. Keeping up with the ever-changing cybersecurity threats in the industrial sector requires a proactive and adaptable approach to the complex job of building secure IIoT and Edge infrastructures. A framework is provided in Figure 9.5 for this purpose. Table 9.5 gives a summary of security techniques for building commercial applications.

9.5.1 EDGE AI AND BLOCKCHAIN FOR SECURE AND TRANSPARENT DATA SHARING

The integration of edge AI and blockchain technologies presents a powerful solution for secure and transparent data sharing in various applications. Edge AI brings processing capabilities closer to the data source, enabling real-time analysis and decision-making. This not only enhances efficiency but also reduces latency and bandwidth usage. Concurrently, blockchain ensures secure and transparent data

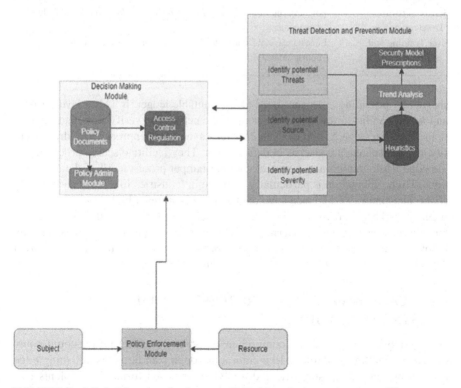

FIGURE 9.5   Minimising security threats in IIoT, AI, and edge environments.[21]

TABLE 9.5
Techniques for Enforcing Security

Security Aspect	Techniques
Data Encryption	Utilise end-to-end encryption protocols such as OPC UA for secure data transmission.
Access Control	Implement strong access control mechanisms to restrict unauthorised system access.
Device Authentication	Deploy robust authentication methods like X.509 certificates for device validation.
Secure Networking Protocols	Choose secure networking protocols such as DDS, LoRaWAN, or Zigbee based on use cases.
Secure Gateway Deployment	Deploy secure gateways to act as intermediaries, enhancing communication security.
OTA Updates	Implement OTA updates using cloud-based platforms like AWS IoT Device Management.
Continuous Monitoring	Employ continuous monitoring tools to detect and respond to security incidents.
Intrusion Detection and Prevention	Integrate intrusion detection and prevention mechanisms for real-time threat mitigation.
Secure Device Management	Implement efficient device management platforms to monitor and control device fleets.
Secure Software Development	Follow secure coding practices and conduct regular security audits during development.
User Training and Awareness	Educate users and stakeholders about security best practices to reduce human-related risks.

sharing by providing a decentralised and immutable ledger. It adds a layer of trust and accountability, allowing participants in the network to verify and audit transactions. Together, edge AI and blockchain offer a synergistic approach to address security and transparency concerns in data sharing. They facilitate secure processing at the edge while maintaining an auditable and tamper-proof record of data transactions. This combination is particularly valuable in industries like healthcare, supply chain, and finance, where sensitive data sharing requires a high level of trust and security.[22] The joint use of edge AI and blockchain not only safeguards against unauthorised access and tampering but also promotes transparency, making it an innovative and robust solution for the evolving landscape of secure and transparent data sharing. The architecture is shown in Figure 9.6.

9.6 SCALABILITY AND FUTURE TRENDS IN IIoT, AI, EDGE COMPUTING

IIoT scalability issues need a comprehensive strategy taking into account many mitigating methods. Methods like embracing open standards, establishing edge computing, and using sophisticated analytics are used to tackle problems like

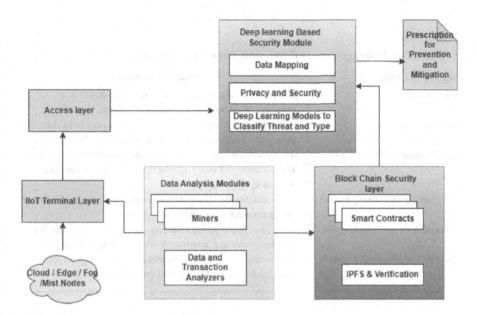

FIGURE 9.6   SecBlock IIoT: Blockchain-enabled security architecture.[23]

data management complexity, network congestion, and interoperability. While optimising algorithms and making use of edge computing resources are necessities for fulfilling processing power needs, scalable storage solutions are essential for ensuring smooth integration with expanding datasets. Strong cybersecurity measures, such as encryption and access restrictions, reduce security and privacy risks.[24]

Furthermore, new developments in edge AI are centred on intelligent data routing, specialised AI chipsets, and decentralised AI processing, all of which improve efficiency and decision-making in real-time. Improved data speeds and ultra-reliable, low latency communication are on the horizon with the advent of 6G networks, which will pave the way for vital IIoT applications. To sum up, a holistic view of scalability takes into account new technologies, established protocols, and long-term planning, while current tendencies point to a world where networked systems are both more intelligent and responsive in the future.[25]

The various security challenges are listed in Table 9.6, which gives a comprehensive idea of where more focus has to be provided to patch the vulnerability of data. The hardware and related scalability issues are summarised in Table 9.7. Reducing hardware costs for making and connecting devices has increased the numbers, connective complexity and evolution in connecting patterns more prominent to attackers. The computing platform, both hardware and software, must cope up with the challenges and provide the functionality to the end user.[26]

The challenges related to data transfer rate are listed in Table 9.8. The low latency communication that is prodigy of 6G communication with seamless, high speed communication rate is also a challenge for the IIoT, AI, and edge applications.[27]

TABLE 9.6

Scalability Challenges for IIoT, AI, and Edge Applications

Scalability Challenges	Mitigation Techniques
Interoperability	Adopt open standards and protocols to ensure seamless communication between devices and platforms.
Network Congestion	Implement edge computing to distribute processing tasks closer to data sources, alleviating congestion and reducing latency.
Data Management Complexity	Employ advanced analytics and machine learning algorithms to derive meaningful insights, reducing the burden on storage and processing resources.
Integration with Growing Datasets	Utilise scalable and cost-effective storage solutions like distributed storage systems and cloud-based options.
Processing Power Requirements	Leverage edge computing resources and optimise algorithms for efficient processing, minimising strain on centralised systems.
Security and Privacy Concerns	Implement robust cybersecurity measures, including encryption and access controls, to safeguard data integrity and user privacy.

TABLE 9.7

Hardware Scalability Challenges for IIoT, AI, and Edge Applications

Emerging Trends in Edge AI Advancements	Description
Decentralised AI Processing	Integration of AI capabilities at the edge of IIoT networks for reduced latency, improved real-time decision-making, and enhanced privacy.
Edge AI Chipsets	Development of specialised AI chipsets for edge computing applications, increasing efficiency and performance for AI tasks at the edge.
Edge-to-Cloud Orchestration	Integration of edge and cloud resources in a complementary and orchestrated manner for optimised resource utilisation.
Intelligent Data Routing	Dynamic routing of data between edge devices and the cloud based on processing requirements enhances efficiency in data transmission.

TABLE 9.8

Data Rate Challenges for IIoT, AI, and Edge Applications

6G and Beyond for Improved Connectivity	Description
Ultra-Reliable Low Latency Communication (URLLC)	Focus on 6G networks, which provide extremely low latency and high reliability for real-time communication in critical IIoT applications.
Enhanced Data Rates	Higher data transfer speeds in 6G networks to support the exponential growth of data generated by IIoT devices.

9.7 STRATEGIES FOR BUILDING A SECURE AND RESILIENT INDUSTRIAL FUTURE

Selecting an IIoT solution involves a comprehensive evaluation across various categories. In terms of device integration, it is crucial to assess compatibility with both legacy and new machines, including support for diverse protocols. Analytics capabilities, including edge analytics and predictive analysis, should be considered for real-time decision-making and early malfunction detection. The total cost of ownership considerations includes time to development, time to market, and pricing structures, with an emphasis on ready-to-use solutions and plugins.[28,29] Security aspects cover edge level security, data security, and application-level security to ensure secure connectivity and user access control. Evaluating the industry presence involves assessing the maturity, size, and partner ecosystem of the IIoT solution provider, supplemented by proof of validations like third-party certificates and customer testimonials. This comprehensive approach aims to address key factors influencing the selection of a robust and suitable IIoT solution, as shown in Table 9.9.

9.8 REGULATORY FRAMEWORKS AND ETHICAL CONSIDERATIONS

Regulatory frameworks and ethical considerations are pivotal elements in the advancement and implementation of technologies, particularly within emerging fields such as AI, data privacy, and the IoT. Regulatory frameworks serve as guidelines and rules established by governments or regulatory bodies to ensure responsible use of technologies, preventing risks to individuals or society. Ethical considerations, conversely, entail an evaluation of the moral implications and societal impacts associated with deploying specific technologies.[40]

Within the realm of AI, regulatory frameworks may address issues like transparency, accountability, bias mitigation, and the responsible integration of AI in decision-making processes. Crucial to this are privacy regulations, such as the General Data Protection Regulation (GDPR), governing the collection, storage, and processing of personal data. Furthermore, ethical considerations in AI encompass concerns related to fairness, accountability, transparency, and potential socio-economic impacts of AI systems.[41,42]

In the domain of IoT, regulatory frameworks concentrate on aspects like data protection, cybersecurity, and standards for interoperability. Ethical considerations for IoT commonly revolve around ensuring user consent for data collection, safeguarding user privacy, and addressing security vulnerabilities to prevent unauthorised access or misuse of connected devices.[28,43]

Both regulatory frameworks and ethical considerations play a pivotal role in shaping the trajectory of technological development and deployment. They provide a structured approach to addressing potential risks and challenges, ensuring that technological progress aligns with societal values and adheres to legal standards. As technology evolves, the ongoing adaptation and enhancement of regulatory frameworks and ethical guidelines becomes increasingly

TABLE 9.9

IIoT Considerations to Form a Cyber-Resilient Strategy

Consideration Category	Consideration Points
Device Integration	• Legacy Systems: Check if the IIoT solution supports integration with older machines or assets.[15] • New Machines: Ensure compatibility with proprietary SDKs and security mechanisms of new machines. • Protocol Support: Verify if the IIoT solution supports required messaging and industrial protocols.[30,31]
Analytics	• Edge Analytics: Assess whether the IIoT solution enables real-time processing at the edge for low latency decision-making. • Predictive Analysis: Evaluate the solution's ability to provide predictive insights for early detection of malfunctions. • KPI Calculation: Check if the IIoT solution offers out-of-the-box KPI services and allows customisation of metrics.[32,33]
Total Cost of Ownership	• Time to Development: Consider the availability of out-of-the-box features and connectors for quick problem resolution. • Time to Market: Evaluate marketplace offerings for ready-to-use solutions and plugins. • Price: Examine different pricing structures, including pay-as-you-go, monthly, or yearly subscriptions, and factor in maintenance and support costs.[34,35]
Security	• Edge Level Security: Confirm if the IIoT solution provides secure edge connectivity for edge analytics. • Data Security: Ensure encrypted data storage and secure data transfer over communication channels. • Application-Level Security: Verify that the solution supports secure access control for managing user permissions effectively.[36,37]
Industry Presence	• Company Maturity and Size: Evaluate the industry presence, vision, and maturity of the IIoT solution provider. • Partner Ecosystem: Consider the strength of the vendor's partner ecosystem for extensions and interoperability. • Proof of Validations: Check for third-party validations, certificates, and customer testimonials as indicators of platform stability and security.[38,39]

imperative to foster innovation while safeguarding individual rights and societal well-being.[44–46]

9.9 CONCLUSION

This chapter explores the transformative potential of Edge, IIoT, and AI for industrial engineering. It highlights the benefits of increased efficiency, better decision-making, and enhanced security while acknowledging challenges like collaboration, continuous improvement, and ethical considerations. The concluding message emphasises the responsibility to harness this technology for a positive future, prioritising both technological advancement and societal well-being.

REFERENCES

1. M. Chu and Y. Song, "Analysis of Network Security and Privacy Security Based on AI in IoT Environment," 2021 IEEE 4th International Conference on Information Systems and Computer Aided Education (ICISCAE), Dalian, China, 2021, pp. 390–393, doi: 10.1109/ICISCAE52414.2021.9590786.
2. N. Bahache and N. Chikouche, "A Comparative Analysis of RFID Authentication Protocols for Healthcare Applications," 2021 International Conference on Artificial Intelligence for Cyber Security Systems and Privacy (AI-CSP), El Oued, Algeria, 2021, pp. 1–6, doi: 10.1109/AI-CSP52968.2021.9671178.
3. J. Shahid, Z. Muhammad, Z. Iqbal, M. S. Khan, Y. Amer and W. Si, "SAT: Integrated Multi-agent Blackbox Security Assessment Tool using Machine Learning," 2022 2nd International Conference on Artificial Intelligence (ICAI), Islamabad, Pakistan, 2022, pp. 105–111, doi: 10.1109/ICAI55435.2022.9773750.
4. M. Kantarcioglu and F. Shaon, "Securing Big Data in the Age of AI," 2019 First IEEE International Conference on Trust, Privacy and Security in Intelligent Systems and Applications (TPS-ISA), Los Angeles, CA, USA, 2019, pp. 218–220, doi: 10.1109/TPS-ISA48467.2019.00035.
5. Y. Zhang, F. Hu, Y. Han, W. Meng, Z. Guo and C. Li, "AI-Based Energy-Saving for Fog Computing-Empowered Data Centers," 2023 International Conference on Mobile Internet, Cloud Computing and Information Security (MICCIS), Nanjing, China, 2023, pp. 16–21, doi: 10.1109/MICCIS58901.2023.00009.
6. F. A. Qazi, "Study of Zero Trust Architecture for Applications and Network Security," 2022 IEEE 19th International Conference on Smart Communities: Improving Quality of Life Using ICT, IoT and AI (HONET), Marietta, GA, USA, 2022, pp. 111–116, doi: 10.1109/HONET56683.2022.10019186.
7. O. Almazrouei, P. Magalingam, M. Kamrul Hasan, M. Almehrzi and A. Alshamsi, "Penetration Testing for IoT Security: The Case Study of a Wireless IP Security CAM," 2023 IEEE 2nd International Conference on AI in Cybersecurity (ICAIC), Houston, TX, USA, 2023, pp. 1–5, doi: 10.1109/ICAIC57335.2023.10044176.
8. N. Moustafa, M. Keshky, E. Debiez and H. Janicke, "Federated TON_IoT Windows Datasets for Evaluating AI-Based Security Applications," 2020 IEEE 19th International Conference on Trust, Security and Privacy in Computing and Communications (TrustCom), Guangzhou, China, 2020, pp. 848–855, doi: 10.1109/TrustCom50675.2020.00114.
9. A. Moumena, "A New Intelligent Jamming Attacks Detection using FCM Clustering Technique Based on Data Mining for Wireless Communication," 2021 International Conference on Artificial Intelligence for Cyber Security Systems and Privacy (AI-CSP), El Oued, Algeria, 2021, pp. 1–7, doi: 10.1109/AI-CSP52968.2021.9671122.
10. Z. Liu, "Construction of Computer Mega Data Security Technology Platform Based on Machine Learning," 2021 IEEE 4th International Conference on Information Systems and Computer Aided Education (ICISCAE), Dalian, China, 2021, pp. 538–541, doi: 10.1109/ICISCAE52414.2021.9590732.
11. N. Mohamed and J. Al-Jaroodi, "Security Applications of Edge Intelligence," 2023 IEEE World AI IoT Congress (AIIoT), Seattle, WA, USA, 2023, pp. 508–512, doi: 10.1109/AIIoT58121.2023.10174600.
12. Y. Oh et al., "Mobilint's ARIES: Chip for Edge AI," 2023 IEEE International Conference on Consumer Electronics-Asia (ICCE-Asia), Busan, Korea, Republic of, 2023, pp. 1–4, doi: 10.1109/ICCE-Asia59966.2023.10326331.
13. L. Zhang, J. Hao, G. Zhao, M. Wen, T. Hai and K. Cao, "Research and Application of AI Services Based on 5G MEC in Smart Grid," 2020 IEEE Computing, Communications and IoT Applications (ComComAp), Beijing, China, 2020, pp. 1–6, doi: 10.1109/ComComAp51192.2020.9398885.

14. G. Czeczot, I. Rojek, D. Mikołajewski and B. Sangho, "AI in IIoT Management of Cybersecurity for Industry 4.0 and Industry 5.0 Purposes," Electronics 12, no. 18 (2023): 3800.

15. G. Dagnaw. "Artificial intelligence towards future industrial opportunities and challenges." (2020).

16. M. O. Ozcan, F. Odaci and I. Ari, "Remote Debugging for Containerized Applications in Edge Computing Environments," 2019 IEEE International Conference on Edge Computing (EDGE), Milan, Italy, 2019, pp. 30–32, doi: 10.1109/EDGE.2019.00021.

17. I. Eugene Berna, K. Vijay, S. Gnanavel and J. Jeyalakshmi, "Impact of Artificial Intelligence and Machine Learning in Cloud Security", in Improving Security, Privacy, and Trust in Cloud Computing, edited by Pawan Kumar Goel, Hari Mohan Pandey, Amit Singhal, and Sanyam Agarwal, 34–58. Hershey, PA: IGI Global, 2024, doi: 10.4018/979-8-3693-1431-9.ch002.

18. M. Santhiya, J. Jeyalakshmi and Harish Venu. "Emerging Networking Technologies for Industry 4.0." in Privacy Preservation and Secured Data Storage in Cloud Computing, edited by Lakshmi D. and Amit Kumar Tyagi, 322–340. Hershey, PA: IGI Global, 2023, doi: 10.4018/979-8-3693-0593-5.ch015

19. Samuel, Prithi, et al. "Artificial Intelligence, Machine Learning, and IoT Architecture to Support Smart Governance." In AI, IoT, and Blockchain Breakthroughs in E-Governance, edited by Kavita Saini, A. Mummoorthy, Roopa Chandrika, N.S. Gowri Ganesh, 95–113. Hershey, PA: IGI Global, 2023. doi: 10.4018/978-1-6684-7697-0.ch007

20. V. Sathya Preiya, V. D. Ambeth Kumar, R. Vijay, K Vijay and N. Kirubakaran, "Blockchain-Based E-Voting System With Face Recognition," Journal of Fusion: Practice and Applications 12, no. 1 (2023): 53–63.

21. https://www.intersecinc.com/blogs/securing-industrial-iot-mitigating-risks-and-ensuring-resilience-in-industry-4-0

22. K. Sha, R. Errabelly, W. Wei, T. A. Yang and Z. Wang, "EdgeSec: Design of an Edge Layer Security Service to Enhance IoT Security," 2017 IEEE 1st International Conference on Fog and Edge Computing (ICFEC), Madrid, Spain, 2017, pp. 81–88, doi: 10.1109/ICFEC.2017.7.

23. A. S. M. Sanwar Hosen, P. K. Sharma, D. Puthal, I. Ho. Ra and G. H. Cho," SECBlock-IIoT: A Secure Blockchain-enabled Edge Computing Framework for Industrial Internet of Things", ASSS '23: Proceedings of the Third International Symposium on Advanced Security on Software and Systems, Melbourne, Australia, July 2023, doi: 10.1145/3591365.3592945

24. Y. Song, S. S. Yau, R. Yu, X. Zhang and G. Xue, "An Approach to QoS-based Task Distribution in Edge Computing Networks for IoT Applications," 2017 IEEE International Conference on Edge Computing (EDGE), Honolulu, HI, USA, 2017, pp. 32–39, doi: 10.1109/IEEE.EDGE.2017.50.

25. Y. Sun, X. Xu, R. Qiang and Q. Yuan, "Research on Security Management and Control of Power Grid Digital Twin Based on Edge Computing," 2021 2nd International Seminar on Artificial Intelligence, Networking and Information Technology (AINIT), Shanghai, China, 2021, pp. 606–610, doi: 10.1109/AINIT54228.2021.00122.

26. P. Gao, R. Yang and X. Gao, "Research on "Cloud-Edge-End" Security Protection System of Internet of Things Based on National Secret Algorithm," 2019 IEEE 4th Advanced Information Technology, Electronic and Automation Control Conference (IAEAC), Chengdu, China, 2019, pp. 1690–1693, doi: 10.1109/IAEAC47372.2019.8997741.

27. M. Caprolu, R. Di Pietro, F. Lombardi and S. Raponi, "Edge Computing Perspectives: Architectures, Technologies, and Open Security Issues," 2019 IEEE International Conference on Edge Computing (EDGE), Milan, Italy, 2019, pp. 116–123, doi: 10.1109/EDGE.2019.00035.

28. D. Ganesh, K. Suresh, M. S. Kumar, K. Balaji and S. Burada, "Improving Security in Edge Computing by using Cognitive Trust Management Model," 2022 International Conference on Edge Computing and Applications (ICECAA), Tamil Nadu, India, 2022, pp. 524–531, doi: 10.1109/ICECAA55415.2022.9936568.

29. Y. Gong, C. Chen, B. Liu, G. Gong, B. Zhou and N. K. Mahato, "Research on the Ubiquitous Electric Power Internet of Things Security Management Based on Edge-Cloud Computing Collaboration Technology," 2019 IEEE Sustainable Power and Energy Conference (iSPEC), Beijing, China, 2019, pp. 1997–2002, doi: 10.1109/iSPEC48194. 2019.8974871.

30. W. Xiao, Y. Miao, G. Fortino, D. Wu, M. Chen and K. Hwang, "Collaborative Cloud-Edge Service Cognition Framework for DNN Configuration Toward Smart IIoT," IEEE Transactions on Industrial Informatics 18, no. 10 (2021): 7038–7047.

31. Vikram Puri, I. Priyadarshini, Raghvendra. Kumar and L. C. Kim. "Blockchain Meets IIoT: An Architecture for Privacy Preservation and Security in IIoT," 2020 International Conference on Computer Science, Engineering and Applications (ICCSEA), 2020, pp. 1–7. IEEE.

32. J. Lee, J. Singh, M. Azamfar and V. Pandhare, "Industrial AI and predictive analytics for smart manufacturing systems", Eds. Masoud Soroush, Michael Baldea, Thomas F. Edgar, Smart Manufacturing, Elsevier, 2020, 213–244, doi: 10.1016/B978-0-12-820027-8.00008-3

33. A. Bhargava, D. Bhargava, P. N. Kumar, G. S. Sajja and S. Ray (2022). Industrial IoT and AI Implementation in Vehicular Logistics and Supply Chain Management for Vehicle Mediated Transportation Systems. International Journal of System Assurance Engineering and Management 13: 673–680. doi: 10.1007/s13198-021-01581-2

34. D. Stadnicka, J. Sęp, R. Amadio, D. Mazzei, M. Tyrovolas, C. Stylios, A. Carreras-Coch, J. A. Merino, T. Żabiński and J. Navarro, "Industrial Needs in the Fields of Artificial Intelligence, Internet of Things and Edge Computing," Sensors 22, no. 12 (2022): 4501.

35. R. Hill and H. Al-Aqrabi, "Edge Intelligence and the Industrial Internet of Things," Advances in Edge Computing: Massive Parallel Processing and Applications 35 (2020): 178.

36. B. Bhandari, "A Hybrid IIoT-Based System for Real-Time Monitoring and Control of Industrial Processes," Journal OF Algebraic Statistics 11, no. 1 (2020): 82–93.

37. Tyagi, Amit Kumar. "Blockchain and Artificial Intelligence for Cyber Security in the Era of Internet of Things and Industrial Internet of Things Applications." In AI and Blockchain Applications in Industrial Robotics, edited by Rajashekhar C. Biradar, et al., 171–199. Hershey, PA: IGI Global, 2024. doi: 10.4018/979-8-3693-0659-8.ch007

38. Pedro Gomes, Naercio Magaia and Nuno Neves. "Industrial and artificial Internet of Things with augmented reality." Convergence of Artificial Intelligence and the Internet of Things (2020): 323–346.

39. Tong Zhang, Yikai Li and CL Philip Chen, "Edge Computing and Its Role in Industrial Internet: Methodologies, Applications, and Future Directions," Information Sciences 557 (2021): 34–65.

40. B. Mao, J. Liu, Y. Wu and N. Kato, "Security and Privacy on 6G Network Edge: A Survey," in IEEE Communications Surveys & Tutorials, vol. 25, no. 2, pp. 1095–1127, Secondquarter 2023, doi: 10.1109/COMST.2023.3244674.

41. Bandar Alotaibi, "A Survey on Industrial Internet of Things Security: Requirements, Attacks, AI-Based Solutions, and Edge Computing Opportunities," Sensors 23, no. 17 (2023): 7470.

42. Sha Zhu, Kaoru Ota and Mianxiong Dong, "Green AI for IIoT: Energy Efficient Intelligent Edge Computing for Industrial Internet of Things," IEEE Transactions on Green Communications and Networking 6, no. 1 (2021): 79–88.

43. Yan Zhao, Ning Hu, Yue Zhao and Zhihan Zhu, "A Secure and Flexible Edge Computing Scheme for AI-Driven Industrial IoT," Cluster Computing 26, no. 1 (2023): 283–301.

44. Tie Qiu, Jiancheng Chi, Xiaobo Zhou, Zhaolong Ning, Mohammed Atiquzzaman and Dapeng Oliver Wu, "Edge Computing in Industrial Internet of Things: Architecture, Advances and Challenges," IEEE Communications Surveys & Tutorials 22, no. 4 (2020): 2462–2488.

45. Pankaj Bhambri and Sita Rani. "Challenges, Opportunities, and the Future of Industrial Engineering with IoT and AI." Integration of AI-Based Manufacturing and Industrial Engineering Systems with the Internet of Things (2024): 1–18.

46. Fatemeh Banaie and Mahdi Hashemzadeh. "Complementing IIoT services through AI: feasibility and suitability." AI-Enabled Threat Detection and Security Analysis for Industrial IoT (2021): 7–19.

10 The Convergence of Medical IoT and Patient Privacy
Challenges and Solutions

Sara Sawant, Yash Mane, Aruna Pavate,
Ashvini Chaudhari and Sweta Dargad

10.1 INTRODUCTION

The study's main purpose is to evaluate the risks and challenges concerning with the application of Internet of Medical Things (IoMT) and identify any existing cyber security vulnerabilities that require attention. In this, the study aims to suggest appropriate approaches that could be adopted for the management of these vulnerabilities. Moreover, the study intends to explore emerging trends in the area of Medical Internet of Things (MIoT) and to identify potential obstacles that might arise in the future. This entails a thorough examination of MIoT's development and how it has affected cybersecurity. The stakeholders stand to gain a great deal from the study's conclusions as they get ready for new developments in the MIoT space.

10.1.1 INTRODUCTION TO MEDICAL IoT

Many parts of the world face formidable challenges in managing their rapidly aging populations, chronic illnesses [1–3], high rates of child mortality, frequent disease outbreaks, harsh living circumstances, low mental health, scarcity of clean drinking water, and growing pollution. Despite the burgeoning demand for medical services, the traditional hospital-based care model persists, wherein patients visit a physician only when they fall ill. Hospitals primarily follow a physician-centered approach that is reactive and neglects patients' active involvement in the medical process [4]. Time constraints, adherence monitoring, an aging population, urbanization, and other challenges pose significant hurdles to hospital-based medical practices. The introduction of IoT technology has significantly changed the healthcare sector and made it possible to offer cutting-edge medical services. Hospitals have traditionally followed a physician-centered treatment strategy that does not include patients as active participants during medical procedures [5]. However, the integration of IoT technology has proven instrumental in proactively monitoring patients' health and providing real-time data to medical staff through IoT-connected medical devices (IoMT). This information is pivotal in preventing device shutdowns and enabling remote

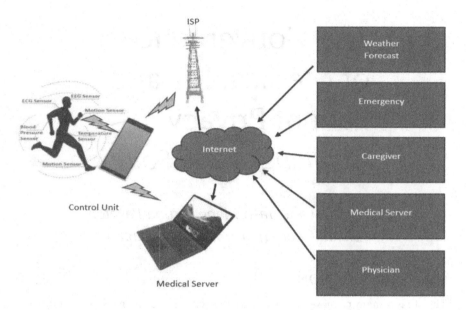

FIGURE 10.1 Usage of MIoT.

monitoring of patients' health [6,7]. The IoMT is a network of medical devices and software that provide direct interaction between medical personnel and their servers. They provide valuable information to humanity by collecting, storing, and process-ing data from various medical devices [8]. The advanced technology also addresses the challenges and obstacles facing hospital-based medical practices, such as limited time, adherence monitoring, an aging population, urbanization, and healthcare work-force shortage. IoMT technology opens up many opportunities for medical services, especially IoMT, and makes managing the healthcare system easier. This technology has revolutionized the healthcare sector, enabling it to offer advanced medical ser-vices, with the potential for further growth and enhancement (Figure 10.1).

10.1.2 HEALTHCARE MONITORING DEVICES

Healthcare monitoring devices have been revolutionized with the introduction of IoT devices. These devices offer numerous opportunities for healthcare professionals to monitor their patients and for patients to monitor their own health as well. In addi-tion, the availability of wearable IoT devices provides a huge range of benefits and challenges for both health officers and their patients.

10.1.2.1 Remote Health Monitoring

IoT-powered remote patient monitoring is a popular healthcare application that lets medical professionals collect critical health metrics from patients who are not physi-cally present in a healthcare facility. The technology automatically gathers patient data, such as heart rate, blood pressure, and temperature, eliminating the need for patients to travel as shown in Figure 10.2. Once the patient data is gathered, algorithms

FIGURE 10.2 Remote health monitoring device.

analyze it and recommend therapy or produce alerts. However, data security and privacy remain a significant challenge in embracing remote patient monitoring devices. With proper security protocols, healthcare providers can confidently use IoT devices to lower expenses, raise the standard of care, and improve patient outcomes [9–11].

10.1.2.2 Glucose Monitoring

Diabetes management is challenging due to the inconvenience and unreliability of traditional manual glucose monitoring methods. IoT devices can offer continuous and automatic monitoring, which solves these challenges. However, designing a small and low-power IoT device for glucose monitoring presents unique challenges. Overcoming these challenges will revolutionize glucose monitoring for patients. IoT devices will play an increasingly crucial role in diabetes management as the technology continues to evolve, providing patients with the tools they need to manage their health effectively. Few of the researchers worked on diabetes management using soft computing concepts without monitoring the glucose [3,12,13]. With the advancement of technology, there might be a possibility of changing the records while training the models [7,14,15].

10.1.2.3 Heart-rate Monitoring

Heart rate monitoring can be challenging for patients in healthcare facilities because of constraints of periodic checks and wired monitoring devices. However, modern IoT devices offer continuous monitoring while allowing patients to move freely. Despite the challenge of achieving ultra-precise results, these devices can provide accuracy rates of 90% or higher [9]. These devices collect data on various physiological parameters and transmit it to healthcare providers for analysis and intervention as shown in Figure 10.3.

10.1.2.4 Hand Hygiene Monitoring

IoT devices can help maintain proper hand hygiene in healthcare facilities, especially hospitals. These devices remind healthcare providers and patients to sanitize

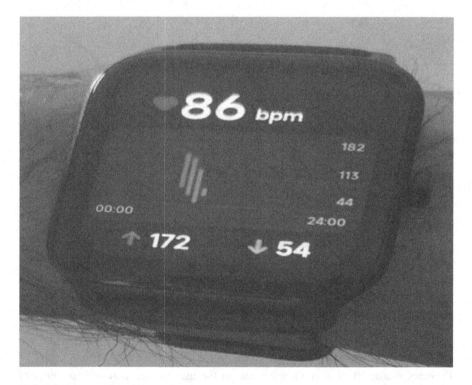

FIGURE 10.3 Heart-rate monitoring device.

FIGURE 10.4 Hand hygiene monitoring device [16].

one's hands before heading into hospital rooms and provide instructions for effective hand sanitation methods. They are only reminders and not a substitute for the actual hand-washing process as shown in Figure 10.4, but studies show that they can reduce hospital infection rates by over 60% [9].

10.1.2.5 Parkinson's Disease Monitoring

IoT sensors continually gather data on Parkinson's symptoms, providing healthcare providers with important insights to adjust treatments and improve outcomes. Patients can continue to be independent using non-intrusive monitoring, permitting them to live their lives outside a hospital setting. The use of IoT sensors is a significant advancement in the treatment of Parkinson's disease, delivering more personalized and effective care while maintaining patients' quality of life.

10.1.2.6 Ingestible Sensors

Ingestible sensors are making it easier to collect data from inside the human body without the requirement for intrusive techniques like inserting cameras or probes.

These sensors can gather relevant information, such as stomach pH levels and internal bleeding sources. They must be easy to swallow and able to dissolve or pass through the human body independently. Numerous institutions are presently engaged in developing such ingestible sensors.

10.1.2.7 Connected Contact Lenses

Smart contact lenses can collect healthcare data without being invasive. They also have micro-picture-taking cameras that use your eyes. This innovation has led companies like Google to patent connected contact lenses. Smart lenses can make your eyes powerful tools for digital interactions, whether for health improvement or other purposes.

10.1.2.8 Connected Inhalers

IoT-connected inhalers as shown in Figure 10.5 can help manage respiratory conditions like asthma or COPD by monitoring attack frequency and environmental triggers. They can also alert patients if they forget their inhaler or use it improperly, reducing the risk of attacks. These inhalers provide valuable insights for patients and data for healthcare providers to ameliorate results.

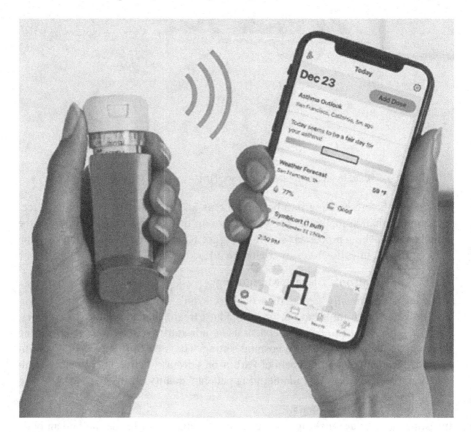

FIGURE 10.5   Connected inhalers.

10.1.2.9 Robotic Surgery

IoT robots are used in the human body to assist surgeons with complex procedures, resulting in less invasive surgeries and faster recovery times. These devices must be small, reliable, and able to interpret complex conditions within the body to make informed decisions during operations. Their increasing use in surgical procedures demonstrates that small-scale robotics technology is being successfully addressed.

10.1.3 Need and Importance of Cybersecurity in the Medical Internet of Things (MIoT)

However, every network-connected device left unprotected in the environment poses a significant security risk. If not addressed, this risk can lead to more severe issues, such as patient health and safety [17]. Notably, IoT and MIoT devices are notoriously difficult to secure, posing a serious security threat. Hackers highly value medical information, which is ten times more valuable to them than a credit card number. Consequently, hospitals are a prime target for cyberattacks. MIoT devices can also provide a backdoor for attackers to access hospital networks and steal confidential data. In one instance, hackers successfully infected blood gas analyzers, vital for monitoring critical care and surgical patients, allowing them to access the hospital's network and steal confidential data. Even with proper precautions in place, the MIoT is always at risk as it sends and receives highly personal data [4]. Unfortunately, the majority of today's systems are not designed to protect such sensitive information. Because of the absence of security protections that most MIoT devices have by design or inadequate security authentication methods that can be readily circumvented by a competent attacker, an attacker can easily eavesdrop on and intercept incoming and outgoing data and information as shown in Figure 10.6. Every level of security concern, including device, system, network, data, access,

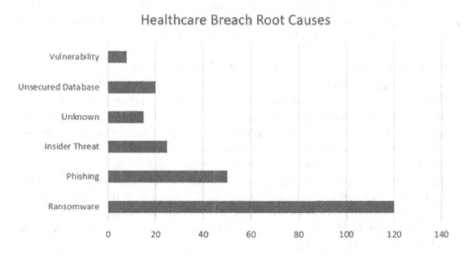

FIGURE 10.6 Core causes of healthcare breaches.

education, and governance, needs to be examined and addressed [18]. The health-care sector is facing more risks to data security and liability as effect of the Internet of Things. IoT raises a lot of security and privacy concerns, but as these devices are automated, the risk is much greater. After gaining access to a network, a hacker can use ransomware to lock down important services or encrypt files until the company pays a ransom. Given how time-sensitive the healthcare sector is, firms frequently have no alternative but to pay the ransom and hold out hope that the money will be eventually retrieved.

10.1.4 CHALLENGES

In order to improve security within the MIoT ecosystem, a few issues that are spe-cific to it must be tackled since they could obstruct the development of ideal security measures. The following list outlines the primary issues with MIoT security and privacy that require immediate attention.

10.1.4.1 Insecure Network

Most Internet of Things (IoT) medical devices mainly rely on wireless networks to communicate, like WI-FI, because they are convenient, widely available, and inexpensive. However, these networks have significant security flaws, like default usernames and passwords that are installed in the device by default and weak authen-tication techniques, which make the device vulnerable to network-level attacks like man-in-the-middle attacks, malware attacks, and WI-FI password cracking [19,20]. The size, complexity, and manufacturer of IoT medical devices generally vary. Most of the devices might have low-speed processors, small amounts of internal memory, and small amounts of storage capacity due to their internal design.

Because most medical IoT devices have limited resources, even a basic dictionary attack can easily exploit and compromise these devices' access control, creating a mega breach in the MIoT health network as a whole. Therefore, it is extremely dif-ficult to develop security solutions that enhance the effectiveness of security while minimizing resource consumption over execution due to the limited resources avail-able and the intricacy of the device manufacturers [5,19,20].

10.1.4.2 Security Patches and Zero-Day Vulnerabilities

Zero-day vulnerabilities are highly likely to be exploited by MIoT devices due to their intrinsic widespread and the speed at which threats are evolving. This raises questions about the necessity of routinely patching possible vulnerabilities on devices in advance before malevolent attackers attempt to exploit them. Conversely, intrud-ers are constantly searching for weak points or connections to take advantage of. For instance, the most vulnerable to security breaches are the antiquated programs that are frequently present in the application layer. Similarly, healthcare system and service providers seldom update physical MIoT devices with the latest firmware, disposing of end-user devices vulnerable to cyberattacks. So as to preserve high availability and avoid zero-day attacks, healthcare firms should update MIoT apps and devices periodically [19].

10.1.4.3 Social Engineering

Due to social media's enormous influence, end users—in this instance, patients—tend to publicly disclose sensitive data on websites like Facebook, Instagram, and others. However, because of their massive user base, cybercriminals see these sites as a fresh and lucrative way to spread malware. Therefore, users should refrain from giving out personal information to random people over the phone or on these websites [19].

10.1.4.4 High Mobility

Many of the devices in the MIoT are typically extremely mobile. For instance, if a patient wears a wearable heart rate monitor that is online, the device will communicate information to the patient's caregivers or the cloud based on the patient's location. When the user is present, it connects to the workplace network; otherwise, it will connect to the home network. Therefore, it is challenging to develop a strong security solution that considers the IoT high degree of mobility because different threat mitigation techniques depend on different environmental security configurations [19,20].

10.2 DISCUSSIONS

10.2.1 Cyberattacks in MIoT

Cyberattacks on MIoT systems present significant risks, including the disruption of medical services as well as the theft of sensitive patient data. Addressing these threats is crucial for maintaining the integrity and safety of healthcare services. Table 10.1 represents the various types of cyberattacks and their impact with levels of impact.

10.2.1.1 Eavesdropping

There are two kinds of eavesdropping attacks: passive eavesdropping and active eavesdropping [4]. Both are primarily focused on gathering information. This is among the easiest approaches attackers are finding to extract data from the sensors. To effectively gather data from hardware devices, attackers must locate and intercept the required hardware [19]. For instance, a patient's vitals could be intercepted during transmission. Such illegally obtained information may be leveraged for a broad variety of attacks. Although encryption can address this issue, it is not always practical, particularly when using devices with low power (Figure 10.7).

10.2.1.2 Attacks Using Reverse Engineering

Typically, an attacker will download the intended software from an app store and assess it locally using a range of tools. A reverse social engineering attack is also known as a person-to-person attack [4]. This makes it possible for the attacker to gain access to the system and obtain information by pretending to be a technician trying to fix a medical system issue at a hospital. Additionally, he might be able to use it to upload malware or locate security flaws.

TABLE 10.1

Types of Cyberattacks Their Impact, and Impact Level

Attack Type	Description	Impact	Impact Level
Eavesdropping [4]	Sensing devices are the main source of information for both passive and aggressive eavesdropping attempts. To get unlawfully obtained data, they intercept hardware, including patient vitals. Even with low-power devices, encryption is not always feasible, although it can be helpful.	Patients' confidentiality might be violated if hackers manage to get private medical data that is sent across devices.	Medium
Attacks Using Reverse Engineering [4]	Reverse engineering assaults might include uploading malware or finding security holes, downloading software from an app store and evaluating it locally, or pretending to be a technician to resolve a medical system problem.	Attackers may find weaknesses in software or medical equipment and use them to their advantage, thereby impairing patient care and causing device malfunction.	High
Attacks by Masquerading [4]	A masquerade attack is a way to access private computer data without authorization by impersonating a person or a network identity. Vulnerability can be caused by insecure authorization processes, especially in the field of healthcare where MIoT devices are essential.	Attackers could take over medical equipment or have illegal access to private patient data.	Medium
Attacks by ZED Sabotage [4]	To be able to break the ZigBee protocol, the developers suggest a ZigBee End-Device attack that would deplete the device's battery through a periodic transmission.	These cyberattacks may interfere with IoT device operation, perhaps resulting in inaccurate data or device malfunction.	Low
Trash Diving Incidents [4]	Dumpster diving is a cyberattack that uses phishing emails to acquire personal and health information from healthcare companies. The information contains patient records, medications, staff names, and other documentation pertaining to waste.	Sensitive patient data may still be on discarded devices, which might result in data breaches.	Low
Traffic Analysis Attacks [4,19]	By taking advantage of network traffic patterns to extract pertinent information, this assault threatens patients' privacy and data confidentiality and may provide hostile attackers the opportunity to purposefully harm medical equipment.	Network traffic analysis may allow hackers to obtain private patient information.	Medium
Modification and Tampering of Messages [4,19]	Intercepting messages Change attacks change application data, including user credentials and product count, between clients and servers to compromise data integrity and can lead medical practitioners to make poor judgments that endanger patient health.	This could interfere with device communication or cause inaccurate data to be transmitted, which might have an effect on patient care.	High

(Continued)

TABLE 10.1 *(Continued)*
Types of Cyberattacks Their Impact, and Impact Level

Attack Type	Description	Impact	Impact Level
Attacks Using Malicious Data Injection [4,19]	When an authorized user blocks other users from sending real messages, an attack by a valid or system authenticated entity might have dangerous consequences in the MIoT system, perhaps resulting in accidents.	This can interfere with IoT device performance, perhaps resulting in inaccurate data or device malfunction.	High
Data Availability Attacks [4,21]	Vital patient health information for hospitals and doctors is compromised when an attacker manipulates signals to interfere with data availability by deleting exchanged messages.	These sorts of attacks have the capacity to take down whole networks, interrupt vital services, and even endanger the lives of patients.	Extreme
Denial of Service (DoS) Attacks [4,19]	DoS attacks compromise medical IoMT systems by blocking access to prescription drugs and medical data, taking advantage of API flaws, blocking real-time data transfer, and interrupting services.	Patient care might be disrupted by these assaults if they cause certain equipment or systems to become unresponsive.	High
Distributed Denial of Service (DDoS) Attacks [4,21]	Attacks from various locations can significantly impact medical systems and devices, potentially affecting patients' lives by making swift reactions difficult.	Attacks from a wide range of sources can have a substantial influence on healthcare systems and equipment, perhaps endangering patients' lives by making quick decisions more challenging.	Extreme
Cross-Site Request Forgery Attacks [4,19]	IoT systems using RESTful APIs are particularly vulnerable to cross-site scripting (CSRF) attacks since insecure applications are secretly used by end users.	Attacks like these might cause equipment to be used for purposes other than intended ones, which could interfere with their ability to operate.	Medium
Pre-Shared Key Attacks [4]	Attackers can quickly access library files because pre-shared keys in Internet of Things (IoT) applications, like the CoAP protocol, are hard-coded.	Hackers may be capable to access networks or devices without authorization, which might result in service interruptions or data breaches.	High
Man-in-the-Middle Attacks [4,19]	Passive and active authentication attacks alter data transmitted between parties with the proper authorization. While active attacks add or modify information without the devices awareness, they interfere with authorized parties' ability to communicate by intercepting communication.	Attackers might intercept and perhaps change device-to-device conversations, resulting in data breaches or the transmission of inaccurate data.	High

(Continued)

TABLE 10.1 *(Continued)*
Types of Cyberattacks Their Impact, and Impact Level

Attack Type	Description	Impact	Impact Level
Blue Bugging Attacks [4]	Attackers can utilize Bluetooth devices to send messages, make phone calls, and access the internet without the user's awareness, a technique known as "bluebugging."	These assaults may result in devices being accessed without authorization, which might cause data breaches.	Low
Rainbow Table Attacks [4]	Salt passwords can offer protection against attacks like "fault and trial" and reverse engineering, which target hash values and passwords.	These assaults may result in illegal access to networks or devices, which can cause data breaches or service interruptions.	High
Ransomware Attacks [4,19]	A common problem with medical equipment such as pacemakers and medicine infusion pumps is IoT ransomware, which encrypts devices and demands a fee to free it. From the perspective of the possibility of death or seriously injured people, this threat is significantly more serious than that of regular malware.	The potential for these assaults to completely disable systems until a ransom is paid, which might seriously impair patient care.	Extreme
Botnet Attacks [4]	In an effort to get private information for malicious purposes, MIoT devices are used in assaults to build bots that provide erroneous patient data and may even launch DDoS attacks.	These assaults have the potential to completely take down networks, drastically impairing patient care.	Extreme
Advanced Persistent Attacks [4]	Deliberate cyberattacks that allow hackers to enter networks without authorization and remain concealed for a long time are known as advanced persistent threats. It is difficult to halt, recognize, or neutralize these assaults utilizing MIoT devices since they monitor network traffic and steal data.	These persistent assaults may result in continuous data breaches and service interruptions.	Extreme

10.2.1.3   Attacks by Masquerading

A masquerade attack is an attempt to gain unauthorized access to private computer data by pretending to be a network identity or another person. If an authorization process is not sufficiently secured, it can become extremely susceptible to a masquerade attack [4]. To gain unauthorized entry into stored confidential health data, an attacker could pose as the authorized user and steal the login credentials for the user's terminal device (such as a smartphone). A hacker can pretend to be a legitimate user in order to access the services of an MIoT device by inserting illegal devices, or the hacker can pretend to be an MIoT device in order to provide customers with fraudulent services. The second scenario, which is especially dangerous for the healthcare sector, is where MIoT devices deliver life-saving services to patients.

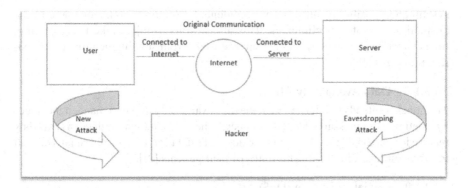

FIGURE 10.7 The eavesdropping attacks in MIoT.

Attacks by ZED Sabotage: The ZED Sabotage attack developers proposed a novel harm to the ZigBee protocol called the ZigBee End-Device attack. Such an attack's primary objective is to vandalize the ZED by waking it up with a periodic signal that drains its battery [4].

10.2.1.4 Trash Diving Incidents
Dumpster diving is the practice of searching trash for personal information and medical records that may be later exploited for hacking purposes [4]. The attack mostly targets big healthcare organizations or businesses, phishing victims primarily by sending phony emails that appear to be from a reliable source. Any medical data recovered during this attack includes staff names, prescription medications, patient records, and other documents and files thrown in the trash.

10.2.1.5 Traffic Analysis Attacks
Patients' privacy and the confidentiality of their data are the main targets of this attack. This is a very dangerous attack where relevant information is extracted by intercepting and analyzing network traffic patterns [19]. This is because it is possible that the activities of MIoT devices could reveal enough details for a malicious adversary to intentionally damage medical devices [4].

10.2.1.6 Modification and Tampering of Messages
The foundation of the message tampering alteration attack is the manipulation of parameters sent between the client and server to change application data, including user credentials and permissions, pricing and product count, and so forth [19]. The intent of the attacker is to undermine the data integrity of the transmitted messages. This happens when a hacker alters messages they receive in order to further their own goals [4]. Consequently, medical professionals may make unwise choices that endanger the health of their patients.

10.2.1.7 Attacks Using Malicious Data Injection
An entity that is either legitimate or capable of system authentication launches this kind of attack [4]. Therefore, this can have hazardous effects on the MIoT

system and possibly result in tragic accidents by sending a false message to the hospital data center or clinicians. This attack aims to inject fake messages into the network in order to prevent authorized users from sending genuine and accurate messages [19].

10.2.1.8 Data Availability Attacks

The attacker intends to interfere with the data availability of the exchanged messages by deleting these messages [4]. This happens when a hacker manipulates signals they receive for personal gain, depriving the hospital's data center or its physicians of vital information regarding the health status of their patients [21].

10.2.1.9 Denial of Service (DoS) Attacks

DoS attacks are accustomed to taking down a particular medical IoMT system or device, stopping nurses and doctors from accessing medical data and information and legitimate patients from getting necessary prescriptions [19]. Transferring and receiving real-time data is prohibited by the disruption and interruption of service. Dos attacks exploit weaknesses in application programming interfaces (API) [4].

10.2.1.10 Distributed Denial of Service (DDoS) Attacks

These attacks can even be launched simultaneously from various nations and geographical locations. The availability of medical systems and devices may be significantly impacted by this, which could have a negative effect on patients' lives by making it more challenging to react swiftly [4]. A notable instance of this attack is when a resource gets flooded with numerous requests [21].

10.2.1.11 Cross-Site Request Forgery Attacks

RESTful API-based IoT systems are more vulnerable to this kind of attack [4]. The end user is tricked into using a vulnerable application without their knowledge by the CSRF technique [19]. If the IoT layer's web interface is not set up correctly, it becomes susceptible to CSRF attacks.

10.2.1.12 Pre-Shared Key Attacks

Pre-shared keys are the foundation of this attack's security mechanism in some IoT applications, like the CoAP protocol [4]. Occasionally, a program uses these keys that are hard-coded. Because of this, the attacker can easily obtain access to the library files if he manages to get his hands on them.

10.2.1.13 Man-in-the-Middle Attacks

This is among the most widespread authentication attacks; it modifies the data sent while controlling and monitoring communication between two authorized parties [19]. This attack may be considered either passive or active. A passive attack is one in which the attacker simply reads and intercepts the messages that are sent back and forth between the two entities [4]. However, an attack is deemed active if the attacker is able to change, manipulate, or add to the sent data or information without the devices' knowledge.

10.2.1.14 Blue Bugging Attacks

In this attack, Bluetooth devices are susceptible to multiple types of attacks, the most serious of which is bluebugging [4]. An adversary could intercept calls, send and receive messages, and connect to the internet on the victim's device without the lawful user's knowledge by taking advantage of vulnerability in outdated device firmware.

10.2.1.15 Rainbow Table Attacks

This attack primarily targets the password and hash value using reverse engineering and a technique called "fault and trial" [4]. Usually, it is a table with passwords and hashes that are run through until a match is discovered. However, using salt passwords may be a great defense against these sorts of attacks.

10.2.1.16 Ransomware Attacks

IoT ransomware is not given enough attention, which could be far more dangerous than traditional malware. Since most of the IoT data is kept in the cloud rather than on individual devices, the conventional ransomware paradigm is simply impractical [4]. IoT ransomware encrypts Internet of Things devices and requests a ransom from the owners to unlock the devices [14,19]. Unfortunately, MIoT devices are frequent targets for ransomware. Certain medical devices, like pacemakers and drug infusion pumps, are locked, which can have catastrophic results because patients may suffer life-threatening injuries or even pass away if these devices are not unlocked in a timely way.

10.2.1.17 Botnet Attacks

These attacks rely on taking advantage of vulnerabilities in MIoT devices to turn them into bots that wait for commands from the adversary through command-and-control to deliver false or inaccurate patient data. They can additionally be implemented in DDoS or DoS attacks to take down the medical system as a whole [4]. Actually, a lot of these attacks aim to reveal private information and use it for nefarious or private gain.

10.2.1.18 Advanced Persistent Attacks

Advanced persistent threats pose a significant security risk to numerous businesses. An advanced persistent threat is a deliberate cyberattack in which a hacker gains unauthorized access to a network and stays hidden for a considerable amount of time [4]. The goal of attackers using advanced persistent threats is to keep an eye on network activity and pilfer important data. It is challenging to stop, identify, or neutralize these types of cyberattacks. Cybercriminals may utilize these MIoT devices to acquire entry to private or business networks. Cybercriminals can use this technique to steal private data.

10.2.2 Security and Privacy Requirements in Medical Internet of Things (MIoT)

10.2.2.1 Localization

Localization is the method of figuring out each wireless device's position within a network. To ascertain whether the sensors are positioned correctly on the body, the former kind of sensor localization is employed. On-body sensor position identification is important for applications like activity recognition. Healthcare systems and medical devices may frequently move both within and outside of network coverage. An intrusion detection method that detects alterations in the network's sensor departure and return times must be implemented in real-time [4].

10.2.2.2 Authentications for MIoT Devices

A device authentication mechanism should be able to create safe, encrypted interactions for data integrity and confidentiality. Any MIoT healthcare system must include device authentication because malicious devices may provide false information about a patient's physical condition that could seriously harm the patient's healthcare decisions [4,21,22].

10.2.2.3 Confidentiality

Data transmissions and medical conversations can be intercepted by an attacker. Patients may be seriously at risk from this eavesdropping technique since the hacker may use the obtained medical information about several illegal activities. Ensuring confidentiality prevents unauthorized personnel from accessing medical data and guarantees that it is only accessible to authorized persons [5,22,23].

10.2.2.4 Data Integrity

In situations where patient safety is at risk, attackers may exploit the broadcast feature of the wireless network to obtain and alter patient data, which could have catastrophic repercussions. makes sure that during transmission or storage, a hostile party cannot try to change or tamper with medical data [4,22,23].

10.2.2.5 Data Availability

Assures that correct data must be accessible to authorized users in order for the right users or nodes to receive timely, dependable entry to the resources. In the MIoT context, services and data provided by medical servers and devices and needed by pertinent users would become unavailable in the case when a denial of service (DoS) attack. Healthcare applications must be accessible at all times to ensure data availability for users and emergency services due to the risk of data loss [4,21,23].

10.2.2.6 Attack Resistance

MIoT systems must be flexible enough to adapt to node failures and prevent only one instance of failure. Additionally, in the event of an attack, the gadgets or the information should be safeguarded by a foundational security schema [4].

10.2.2.7 Tamper-Proof MIoT Devices

The physical theft of MIoT devices, especially ambient sensors, can reveal security information to hackers. The tamper-resistant integrated circuits (TRICs) that prevent outside parties from deciphering the codes placed on the devices once they are installed should be a minimum feature of the MIoT medical devices included in the systems [4].

10.2.2.8   Data Usability

Data usability makes sure that the data is only accessible to individuals who are authorized [4,21].

10.2.2.9   Data Auditing

Monitoring and analyzing access to medical records is a standard procedure for identifying and tracking unusual or suspicious events, as well as a crucial method of managing the utilization of resources [4,5,22].

10.2.3   SOLUTIONS FOR CYBERATTACKS ON MIoT

10.2.3.1   Access Control

Access control outlines who is permitted access to MIoT devices and medical data, as well as the proper degree of access that should be granted. As a result, it must authenticate the person trying to access the data (for example, by using a password, and fingerprint). In order to guarantee optimal security and privacy, well-designed access control needs to be established for both IoT healthcare applications and devices. However, when it comes to physical security, which is another facet of access control, MIoT devices, and medical data should be safeguarded against physical theft, mishaps, environmental dangers, and sabotage whenever necessary. All security and privacy regulations must also be followed [19,24].

10.2.3.2   Data Encryption

Data is protected when transferring or storing data thanks to data encryption. Even in the unlikely event that an attacker gains entry to the medical transmission or database media, strong data encryption might present challenges for them to read private health information [19,24].

10.2.3.3   Data Auditing

When attempting to identify the origin of a security breach, audits are highly helpful as they enable the examination of underlying data (such as alerts from the system, network activity, and user access). However, because it extends to the cloud, the medical IoT network cannot be completely trusted. Therefore, to be able to detect disturbances and irregularities occurring throughout the cloud network, a system of auditing is imperative [19].

10.2.3.4   Data Reduction

According to the concept of "data minimization," MIoT services should only collect personal health information (PHI) when absolutely necessary, and they ought to only keep the information for the duration that is required to achieve the goals of the requested services by users. Reducing the total quantity of personal patient data collected is among the most popular and effective data minimization strategies in the healthcare industry. However, only gathering the necessary and pertinent amount of patient data in accordance with the intended purpose; erasing or hiding personal data that is no longer needed or obsolete; and carrying out routine assessments to guarantee the relevance and sufficiency of the data [19].

10.2.3.5 Devices for Inventory

Healthcare assets that organizations cannot see, they cannot safeguard; hence, a complete inventory of all organizational assets is necessary. Risk assessments must be carried out on a frequent basis in order to detect possible problems, as many IoT devices are added without a comprehensive risk assessment. Some manufacturers offer inventory solutions that can identify IoT devices on a network without interfering with their functionality [19].

10.2.3.6 Use the Best Practices

Improving and securing a standard MIoT environment requires adhering to best practices, which include staying away from hard-coded passwords., putting firewalls and honeypots in place to attract and neutralize attackers, and encrypting sensitive data. However, because these IoT devices and applications are constantly connected to networks, it is important to maintain network security through the use of intrusion prevention systems (IPSs), intrusion detection systems (IDSs), SSL/TSL security protocols, and hypertext transfer protocol secure (HTTPS) communication mechanisms. Furthermore, a risk assessment needs to be completed before any devices are deployed in order to find any weaknesses before configuring the environment. Software and hardware also need to be updated on a regular basis [19].

10.2.3.7 Increased Consciousness

Continuous medical staff training is crucial to enhancing patient safety and well-being, and employers within the medical field should generally be cognizant of information security principles in order to give MIoT applications the security and protection they require and private patient data. Employee education should include imparting sufficient knowledge on patient rights, privacy, and data security [19,24].

10.2.3.8 Upgrade Our Software

Verify that the operating system and all software are up-to-date. These updates contain necessary patches that prevent would-be thieves from taking advantage of previously identified software vulnerabilities. If you do not apply the necessary software updates, criminals can still exploit the flaws in earlier versions. Cybersecurity is just one of the many things you will lose out on if you choose to ignore those software update alerts [4].

10.2.4 Future Directions for Protection of MIoT

In summary, by providing effective services on time, medical IoT-based innovations facilitate the improvement of healthcare facilities' efficiency and offer a plethora of helpful services and applications to enhance the quality of medical care [19]. We've outlined some of the major developments and trends in medical IoT security that we can look forward to in the coming years.

10.2.4.1 Blockchain Technology

10.2.4.1.1 Blockchain-Enabled Decentralized Security

Using blockchain technology, implement a decentralized security architecture to disperse security protocols among cloud systems and MIoT devices. Because blockchain technology cannot be altered, ransomware assaults find it more difficult to damage the system as a whole and to breach individual security measures [4,25,26].

10.2.4.1.2 Smart Contracts for Secure Transactions

Smart contracts on blockchains can be used to automate safe transactions between cloud systems and IoT devices. This lessens the possibility that ransomware demands will be fulfilled by guaranteeing that transactions are authentic and impervious to tampering [4].

10.2.4.1.3 Blockchain-Based Identity Management

Blockchain integration is needed to maintain MIoT device identities securely. Because every device's identification is recorded on the blockchain, illegal gadgets cannot be converted into bots. The blockchain sends out notifications in response to any effort to change the device's identification [25,26].

10.2.4.1.4 Blockchain Consortiums

Create blockchain alliances in the medical field to set MIoT cybersecurity best practices and standards. Work together to create shared blockchain networks that improve healthcare ecosystem security overall [4,26].

10.2.4.2 AI-Powered Anomaly Detection

An effective technique for improving the cybersecurity of medical IoT systems is anomaly detection driven by AI. Machine learning algorithms are able to examine data patterns and spot anomalies or departures from typical behavior [27]. There are many parameters considered for MIoT security as shown in Figure 10.8. Here's how to use anomaly detection driven by AI for medical IoT cybersecurity [27–29].

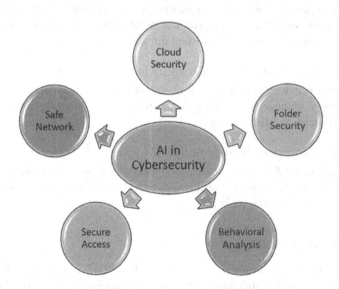

FIGURE 10.8 Several AI features for medical IoT security.

10.2.4.2.1   Behavioral Analysis

Develop machine learning models to comprehend the typical usage, data transfer rates, and communication patterns of medical IoT devices. Any departures from these well-established patterns might point to a possible security risk.

10.2.4.2.2   Continuous Learning

Put in place mechanisms for ongoing learning so the anomaly detection system can adjust to environmental changes. This is particularly crucial in dynamic healthcare environments where software updates and new devices are introduced frequently.

10.2.4.2.3   Attack Pattern Recognition

Teach the AI model to identify particular attack patterns that are frequently connected to cyber threats in the medical field. Such patterns may point to ransomware, malware, or unwanted access attempts.

10.2.4.2.4   Data Encryption and Anomaly Detection

To make sure that the intercepted data is safe and unreadable even in the event that an abnormality is found, combine anomaly detection with data encryption.

10.2.4.3   Zero Trust Security Model

10.2.4.3.1   Zero Trust Access Controls

Make sure that all users, devices, and systems are consistently authorized and authenticated by implementing Zero Trust access controls for MIoT devices. This lessens the possibility of ransomware spreading and unwanted access.

10.2.4.3.2   Continuous Monitoring and Analytics

Include analytics and continuous monitoring in a Zero Trust framework. To ensure ongoing monitoring, this entails analyzing MIoT network behavior in real-time to look for indicators of advanced persistent threats.

10.2.4.3.3   Zero Trust Privilege Management

Use the least privilege principle to limit access permission by implementing Zero Trust privilege management. This reduces the possibility that hackers will obtain continuous access to MIoT systems.

10.2.4.3.4   Continuous Security Audits

In keeping with the Zero Trust model, carry out ongoing security audits. To find and fix vulnerabilities, evaluate the security posture of networks, access controls, and MIoT devices on a regular basis [30].

10.3   CONCLUSION

For safeguarding Healthcare's Future within the MIoT Era, strong cybersecurity is essential for MIoT applications since it holds the key to improved healthcare. This concludes with a call to action. The use of proactive cybersecurity measures becomes crucial as MIoT transforms the healthcare industry. It's not only about data security; additionally, it is about patient privacy protection, device dependability, and strengthening the foundation of healthcare systems. The conclusion reached is a simple directive: embrace cybersecurity in MIoT and enforce it. The success of cybersecurity in the MIoT is critical to the future of healthcare, and the conclusion calls for a steadfast dedication to managing this junction with accuracy, creativity, and a constant focus on safe, patient-centered care. Our review also highlights the importance of the MIoT ecosystem and the contemporary threat environment. Numerous studies have suggested new integrations and enhancements to improve MIoT security, but they do not address the underlying issue of inadequate security within systems, which leaves a gap in security and privacy. We have realized that, in order for security to be effective, it must be integrated rather than patched. It must be a crucial component for the widespread use of MIoT ecosystem. Although MIoT security has attracted a lot of interest, with pertinent standards and technical requirements are still developing and have not yet reached the ideal stage of maturity. Currently, a large number of researchers are working toward innovative, secure MIoT solutions that will provide a wide range of beneficial services and applications to enhance the effectiveness of health care while always protecting privacy and security. This is due to the recent demands for the security and privacy of the MIoT. In the long term, this would be driven by the integration of blockchain, artificial intelligence, critical secure design, embedded security, and strict laws and regulations. In conclusion, this study offers a comprehensive summary of security and privacy issues in the medical IoT, along with actionable countermeasures and solutions, important challenges, and future directions.

REFERENCES

1. A. Pavate, N. Ansari, "Risk Prediction of Disease Complications in Type 2 Diabetes Patients Using Soft Computing Techniques," *2015 Fifth International Conference on Advances in Computing and Communications (ICACC)*, Kochi, India, 2015, pp. 371–375, doi: 10.1109/ICACC.2015.61.
2. A. Pavate, P. Nerurkar, N. Ansari, R. Bansode, Early Prediction of Five Major Complications Ascends in Diabetes Mellitus Using Fuzzy Logic. In: Nayak, J., Abraham, A., Krishna, B., Chandra Sekhar, G., Das, A. (eds) Soft Computing in Data Analytics. Advances in Intelligent Systems and Computing, vol 758. Springer, Singapore, 2019.
3. A. Pavate., et al. "Diabetic Retinopathy Detection-MobileNet Binary Classifier". Acta Scientific Medical Sciences, 4.12 (2020): 86–91.
4. Y. Perwej, N. Akhtar, N. Kulshrestha, P. Mishra, "A Methodical Analysis of Medical Internet of Things (MIoT) Security and Privacy in Current and Future Trends", Journal of Emerging Technologies and Innovative Research, 9(1):346–371, 2022.
5. W. Sun, Y. Li, F. Liu, S. Fang, G. Wang, "Security and Privacy in the Medical Internet of Things: A Review", Hindawi Security and Communication Networks, 2018, 5978636, 2018.
6. M. Haghi, K. Thurow, Dr. I. Habil, R. Stoll, Dr. M. Habil, "Wearable Devices in Medical Internet of Things: Scientific Research and Commercially Available Devices", Healthcare Informatics Research, 23(1):4–15, 2017.
7. P.N. Srinivasu, N. Norwawi, S.S. Amiripalli, P. Deepalakshmi, "Secured Compression for 2D Medical Images Through the Manifold and Fuzzy Trapezoidal Correlation

Function", Gazi University Journal of Science, 35.4 (2022), 1372–1391. https://doi.org/10.35378/gujs.884880

8. J. Tan, "Cybersecurity in the Internet of Medical Things", New York University Journal, 10(3):49–58, 2020.

9. https://ordr.net/article/iot-healthcare-examples/ [accessed Nov 21, 2023]

10. https://www.hfmmagazine.com/articles/3760-hand-hygiene-device-makers-ex-pect-growth [accessed Dec 4, 2023]

11. https://www.airistaflow.com/solutions/healthcare/hand-hygiene/ [accessed Dec 4, 2023]

12. A. Pavate, R. Bansode, Design and Analysis of Adversarial Samples in Safety–Critical Environment: Disease Prediction System. In: Gupta, M., Ghatak, S., Gupta, A., Mukherjee, A.L. (eds) Artificial Intelligence on Medical Data. Lecture Notes in Computational Vision and Biomechanics, vol 37. Springer, Singapore, 2023.

13. A.A. Pavate, R. Bansode, "Analyzing Probabilistic Adversarial Samples to At- tack Cloud Vision Image Classifier Service," *2021 International Conference on Data Analytics for Business and Industry (ICDABI)*, Sakheer, Bahrain, 2021, pp. 689–693, doi: 10.1109/ICDABI53623.2021.9655806.

14. R.V.S.S. Lalitha, P.N. Srinivasu (2017). An Efficient Data Encryption Through Image via Prime Order Symmetric Key and Bit Shuffle Technique. In: Satapathy, S., Bhateja, V., Raju, K., Janakiramaiah, B. (eds) Computer Communication, Networking and Internet Security. Lecture Notes in Networks and Systems, vol 5. Springer, Singapore. https://doi.org/10.1007/978-981-10-3226-4_26

15. N. Dey, A.S. Ashour, S.J. Fong, C. Bhatt, "Wearable and Implantable Medical Devices: Applications and Challenges" (ISSN Book 7) 1st Edition, Kindle Edition ISBN-13 978-0128153697, Publisher Academic Press, Edition 1, 6 September 2019.

16. https://propellerhealth.com/press/press-releases/asthma-patients-maintain-high-medication-adherence-when-using-propeller-health-to-manage-their-condition/ [accessed Dec 4, 2023]

17. M. Ali, "Internet of Things Medical Devices Cybersecurity", Journal of Iowa State University, 2(1):20–25, 2017.

18. G.W. Jackson Jr, S.M. Rahman, "Exploring Challenges and Opportunities in Cybersecurity Risk and Threat Communications Related to the Medical Internet of Things (Miot)", International Journal of Network Security & Its Applications, 11:83–86, 2019.

19. M. Elhoseny, N.N. Thilakarathne, M.I. Alghamdi, R.K. Mahendran, A.A. Gardezi, H. Weerasinghe, "Security and Privacy Issues in Medical Internet of Things: Over-View, Countermeasures, Challenges and Future Directions", Sustainability, 13(21):8–11, 2021.

20. S. Selvaraj, S. Sundaravaradhan, "Challenges and Opportunities in IoT Healthcare Systems: A Systematic Review", Springer Nature Switzerland AG, 20:8–9, 2019.

21. N.M. Thomasian, E.Y. Adashi, "Cybersecurity in the Internet of Medical Things", Health Policy and Technology, 10(3):51–57, 2021.

22. S. Nasiri, F. Sadoughi, M.H. Tadayon, A. Dehnad, "Security Requirements of Internet of Things-Based Healthcare System: A Survey Study", Acta Informatica Medica, 27(4):253–258, 2019.

23. F. Kamalov, B. Pourghebleh, M. Gheisari, Y. Liu, S. Moussa, "Internet of Medical Things Privacy and Security: Challenges, Solutions, and Future Trends from a New Perspective", Sustainability, 15(4):22, 2023.

24. E. Fazeldehkordi, O. Owe, J. Noll, "Security and Privacy in IoT Systems: A Case Study of Healthcare Products", IEEE Journal, 13:6–8, 2019.

25. M.A. Rahman, M.S. Hossain, M.S. Islam, N.A. Alrajeh, G. Muhammad, "Secure and Provenance Enhanced Internet of Health Things Framework: A Blockchain Managed Federated Learning Approach", IEEE Access, 8:14–17, 2020.

26. B. Pradhan, S. Bhattacharyya, K. Pal, "IoT-Based Applications in Healthcare Devices", Hindawi Journal of Healthcare Engineering, 2021, 6632599, 2021
27. B. Upadhay, "Securing IoT using AI", EDUZONE: International Peer Re- viewed/ Refereed Multidisciplinary Journal (EIPRMJ), 2022.
28. D. Lee, "Application of Artificial Intelligence-Based Technologies in the Healthcare Industry: Opportunities and Challenges", *International Journal of Environmental Research and Public Health*, 18(1), 271, 2021
29. I. Keshta, "AI-Driven IoT for Smart Health Care: Security and Privacy Issue", Informatics in Medicine Unlocked, 30:7–8, 2022.
30. "The Right Approach to Zero Trust Security for Medical IoT Devices", Whitepaper Paloalto Networks, 2022.

11 Toward a Trusted Smart City Ecosystem

IoE and Blockchain-Enabled Cognitive Frameworks for Shared Business Services

Puja Das, Chitra Jain, Ansul, and Moutushi Singh

11.1 INTRODUCTION

The building of smart cities is becoming more difficult due to growing urban populations, increased traffic, rising energy consumption demands, and new regulations about health, education, and other areas. How can we incorporate new ideas and technologies, such as large-scale data, the Internet of Everything (IoE), and green connectivity, to ensure that emerging technologies drive the development of smart cities?

The development of smart cities is based on massive information technology, which also serves as a crucial assurance for the proper creation and execution of administration and development plans. High standards for large-scale data solutions are set by the natural combination of innovative governance, intelligent transportation, smart communication, intelligent medical services, and other amenities inside a smart city, all based on autonomy and precise data. Creating or maintaining an extensive information system with outstanding performance and dependability, many uses, anti-attack protection, and recovery from disaster capabilities comes at a considerable cost. Furthermore, blockchain technology can address every issue with the centralized database, including its lack of adaptability and transparency. The problem for future smart cities will come from the confluence of both developments since the Internet of Things (IoT) gadgets and the general public will produce enormous amounts of data daily [1]. These facts must be assimilated, thought through, and replied to securely and thoughtfully. At the smartphone's edge, the IoE, blockchain, and machine intelligence combine to offer innovative city-sharing economy solutions that allow both parties to interact in a fully distributed way without the need for a third-party middleman.

The goal of the energy internet is to increase the adaptability and effectiveness of the electricity system by facilitating the broad connection of dispersed sources of clean energy. However, when resolving issues with user credit establishment, centralized information upkeep, and user confidentiality disclosure, the electricity system still has certain technological flaws [2, 3]. The electricity sector's underlying

DOI: 10.1201/9781003466284-11

market structure and company structure, particularly in the power industry, have remained relatively the same by the energy internet. In addition to being a straight origin of a massive number of facts, the IoE technology is a crucial terminal tentacle in the green internet. As the IoE, technology becomes increasingly prevalent and integrated into every facet of our lives, we are ushering in a revolutionary era where everything is interconnected. This is not just a change, but the dawn of an exciting new epoch in human society. Thousands of new machinery pieces, including wearables, smartphones, connected houses, intelligent environments, and more, will be linked to the internet. This will present both opportunities and challenges for establishing the IoE, including identifying identity, privacy breaches involving data leaks, high costs associated with developing and maintaining databases, and challenges in maintaining and upgrading equipment [4]. The foundation of Bitcoin is the technology known as blockchain. It is a co-trust intelligent book innovation that uses encryption as its foundation.

When blockchain appeared in Bitcoin, the banking sector was the initial group to take an interest in it. Later, the system was expanded to include smart contracts, allowing blockchain to be utilized in various fields. As a result, blockchain applications have attracted widespread attention outside the financial industry [5]. The transportation, accountability, equality, and collaborative features of blockchains can become a significant technical answer to the new environmentally friendly issues associated with the forthcoming energy of the web and significantly influence technological advancement and industry changes in the decades to come. Table 11.1 represents the abbreviation used in this chapter.

11.1.1 BLOCKCHAIN TECHNOLOGY

Any valuable item expressed in code, including properties, copyrights, voting records, accounting books, and other assets, can be recorded in a blockchain, a cryptographic chain. Every entity needs its ledger in this era of connectivity, which is why blockchain is a suitable answer. Blockchain is positioned as the next generation of a worldwide certification and exchange protocol, diverging from the traditional

TABLE 11.1

Acronyms and Their Concise Abbreviations

Symbol	Abbreviation
IoE	Internet of Everything
P2P	Peer-to-peer
D2D	Device to device
DNN	Deep Neural Network
OBCU	On-board CheckUp
GPS	
NBT	Naïve Bayes Theorem
CNN	Convolutional Neural Network
MT	Markel Tree

internet, which primarily transmits information. Blockchain is designed to transport value across networks [6,7]. It facilitates trust between two untrusted parties without requiring third-party endorsements and significantly lowers credit costs.

Blockchain initially emerged as an open and unrestricted public blockchain ledger, exemplified by BTC and ETH. Nevertheless, a variety of adaptations have since emerged to address different requirements and degrees of decentralization. This expands the potential applications of blockchain, reaching industries beyond financial technology (Fintech) and extending into various organizations. Blockchain categories are categorized according to parameters like view, modified access to the ledger, involvement in the verification and consensus mechanism, and the extent of distribution. In a public blockchain, the record is accessible to everyone, allowing unrestricted viewing. Conversely, in a private blockchain, permission to view or modify is restricted to only a few participants [8]. In a regulated ledger, recognized members possess the authority to record transactions and contribute to the consensus mechanism. Conversely, an unrestricted ledger allows anyone to document transactions and engage in the decision-making process. However, a regulated blockchain is overseen by a solitary entity. A consortium ledger represents an adapted version of the regulated blockchain, encompassing many entities. In consortium blockchain, identified participants are empowered to record transactions, while a select subcategory of these participants can verify transactions and contribute to the agreement process. Table 11.2 depicts a comparative analysis between different type of existing consensus algorithms.

Choosing the appropriate blockchain framework is contingent on the characteristics of the planned application and its defined features. Essential elements to take into account include:

- **Confidentiality:** Confidentiality in the context of blockchain refers to the level of anonymity a participant maintains within the network. It pertains to whether the participant's real identity can be traced on-chain. Different blockchains offer varying degrees of anonymity [9]. Cryptocurrencies like Bitcoin offer complete on-chain pseudo-anonymity on public blockchains. However, private blockchain solutions, such as Hyperledger Fabric, often require known identities, especially in enterprise settings.
- **Scalability:** The scalability of different blockchain networks varies. A few participants can be accommodated using Hyperledger Fabric as a platform. Ethereum, however, is very scalable. The efficiency, expressed in terms of TPS, also affects scalability. The blockchain network's consensus system and the method used to validate freshly created blocks are closely linked to the network's sustainability.
- **Contribution:** How does one go about becoming a member of the network? Multichain and Hyperledger are examples of permissioned blockchain systems that need users' prior consent to join the network. On the other hand, public and permissionless blockchain systems allow anyone to create a public/private key pair and join the network.
- **Contact and Confidentiality Strategy**: The choice of ledger frameworks is influenced by the constraints on information extraction and transparency

TABLE 11.2

Comparative Analysis of Various Consensus Algorithms

Consensus Mechanism	Characteristics	Use Cases	Advantages	Disadvantages
Proof of Work (PoW)	– Requires miners to solve complex mathematical problems.	– Bitcoin and Ethereum.	– Security through computational work.	– Energy-intensive.
Proof of Stake (PoS)	– Validators are chosen based on the amount of cryptocurrency held.	– Cardano, Tezos.	– Energy-efficient.	– Potential for centralization.
Delegated PoS (DPoS)	– Users vote for a set number of delegates to validate blocks.	– EOS, TRON.	– High throughput.	– Relies on elected delegates.
Practical Byzantine Fault Tolerance (PBFT)	– Network participants agree on a single state of the system.	– Hyperledger Fabric.	– Low latency.	– Limited scalability for large networks.
Proof of Authority (PoA)	– Validators are approved by a central authority.	– Quorum blockchain.	– High throughput.	– Centralized control.
Proof of Burn (PoB)	– Participants show proof of "burning" coins, making them unusable.	– Slimcoin.	– Token distribution without mining.	– Coins are permanently destroyed.
Proof of Space (PoSpace)	– Participants allocate storage space for mining.	– Chia Network.	– Energy-efficient, utilizes storage space.	– Potential for storage space centralization.
Proof of Elapsed Time (PoET)	– Participants wait for a randomly chosen time period.	– Hyperledger Sawtooth.	– Energy-efficient.	– Requires a trusted execution environment.

among involved parties. Various blockchain ledgers provide different levels of confidentiality and transparency. Permissioned blockchains restrict ledger access to pre-registered and confirmed participants [10]. Conversely, public blockchain ledgers allow everyone to observe transaction networks through resources like Block Explorer. In a blockchain ledger with open access, entry to the decentralized ledger is facilitated by generating a public and private key combination after the sign-up process.

Three generations of blockchain are distinguished by recognized academic Melanie Swan: Blockchain technology has evolved through distinct phases, often referred to as stages 1.0, 2.0, and 3.0. The initial phase, Blockchain 1.0, primarily focused on the introduction and utilization of Bitcoins and other cryptocurrency. A more comprehensive range of commercial and promotional activities, and commercial applications—such as those involving debt, insurance companies, stocks, and smart contracts—are covered by Blockchain 2.0. Blockchain 3.0 affects the administration, medical care, research, civilization, and artistic endeavors in addition to the commercial and commercial spheres, promoting decentralized and cooperative communities [10,11]. Hashing, timestamps, asymmetric encryption, Merkle trees (MTs), proof of work, and peer-to-peer (P2P) networking are just a few of the various technologies that blockchain combines. Although these technologies are not brand-new, blockchain cleverly blends them with cutting-edge ideas to create a distinct system design and working procedure. Particularly, blockchain ensures consistency and network trust by securely recording every bit of transaction information.

11.1.2 HASH

Hashing functions provide hash values by converting a variety of length inputs to a predetermined set of outputs using a hash algorithm. Hashing may be used for conflict resistance and puzzle friendliness, two qualities that are essential for cryptography security. By using asymmetric encryption, timestamps, MTs, and other complex cryptographic procedures, these advanced approaches are essential for improving safety precautions. Businesses may guarantee the protection of sensitive data and information integrity against potential cyber-attacks by implementing these cutting-edge techniques [12].

11.1.3 ASYMMETRIC CRYPTOGRAPHY

In this method, two different functions are needed to encode and decode data using asymmetric encryption. The public key and the private key are the keys. After matching the private key, we may be able to interpret data encoded with the public key, and vice versa. It guarantees that the blockchain meets ownership verification and security criteria. RSA, DSE, Elliptic Curve Cryptography (ECC), and other algorithms are standard.

11.1.4 TIMESTAMP

The existence of specified data at a given moment can be demonstrated using the timestamp server. The timestamp will be added to other data in the block, and its random hash will be used to create a chain. Another block will then join this block, and so on.

11.1.5 THE CONSENSUS APPROACH

The mechanism for consensus is a collection of rules put in place by distributed ledgers to guarantee consistency between the records kept by various nodes within a given time. Depending on the needs of the business and performance factors, typical approaches include delegated proof of stake (DPoS), proof of work (PoW), proof of

TABLE 11.3

A Comparative Analysis of Types of Blockchain

Parameter	Public	Private	Consortium
Permission to read	None	Specific operation	Member of consortium
Permission less	High	Very low	low
Delay in network	Low	High	High
Throughput	Low	High	High
Main transaction	None	Specific operation	Member of consortium
Permission to write	Every node	Specific Authority	Consortium member
Permission to block creation	Every node	Operator	Consortium member
Consensus	PoW	BFT	BFT, PoS
Immutable	Yes	Yes	Yes
Example	Bitcoin	Ethereum	Hyper ledger

stake (PoS), and other Byzantine fault tolerance algorithms [13]. PoW encourages user involvement, and the system becomes more secure as more people participate. However, its high energy consumption and propensity for power concentration present difficulties. Instead of pure PoS systems, PoS is more energy-efficient. Still, it also carries more centralization concerns and a weaker credit foundation because it depends on users' holdings and the length of ownership (currency age). In DPoS, 101 representatives are chosen by shareholders by voting, creating super nodes or pools with equal rights [14]. Table 11.3 represents a comparison between all types of blockchains. These delegates make sure that everyone agrees. Consensus techniques transfer confidence from people and institutions to machine-based algorithms by enabling reliable data exchange across all nodes without depending on third-party trust affirmations. This autonomy embodies the fundamental principles of blockchain technology, which forbid any third-party from interfering with the system's proper operation.

11.1.6 PEER-2-PEER SYSTEM

The P2P system is the foundation for the blockchain, which is distributed using flood algorithms. The longest chain is always considered accurate by the node. In the event of divergence, a reserve chain is formed while the subsequent block is being looked for. Only confirmation from some nodes is needed to record a transaction, provided that enough nodes have sufficient information. You may download the deleted node at another point in time.

11.1.7 CHAIN STRUCTURE AND MERKLE TREE

Large-scale data can be securely verified through the MT. In this binary tree structure, each end node represents a hash of a data block. The identification of a non-end node is achieved by concatenating the hashes of its two forks. The Merkle Root (MR) is the origin block of the MR created in this manner. Until the MR, all modifications

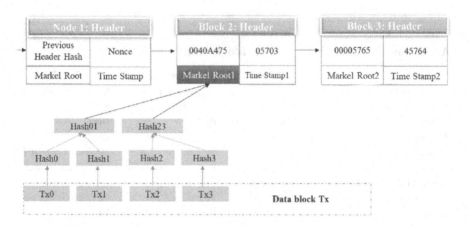

FIGURE 11.1   The basic structure of blockchain.

made to the data blocks are reflected. One hash-valued coding used in blockchain technology is called MR. Certain nodes in a blockchain use simplified payment verification (SPV) to confirm specific transactions within a block without copying the complete block [14]. By making use of the hierarchical structure of the MT, SPV allows fractional blocks to copy only a portion of a block to verify and authenticate the transactions that pertain to them. Because the transactions in a block interlock with one another through the MR and the hash info is kept by loading the hash info of the former block, as depicted in Figure 11.1, the linked construction of the node and the MR makes it hard to break [13]. It is simple to identify the connection structure of a blockchain inference, which allows for blockchain operation manipulation.

While 51% of attackers are expensive, blockchain technology is difficult to breach. The blockchain is highly safe, with strong computation support or a solid proof of workload. The attacker's sole option is to modify their agreement and attempt to recover the funds from their most recent expenditure [15]. Their solitary goal in completing the deal is to further their interests—fraudulent transactions and double-flower difficulties. As a result of the massive difference between contributions and productions, the information on the decentralized system is supposedly unchangeable.

This study provides insightful information for academics and application designers examining the relationship between blockchain technology and smart cities. Among its contributions are:

1. Outlining the salient features of blockchain technology.
2. List and discuss the necessary conditions to use blockchain technology in smart cities.
3. Putting out a blockchain-based conceptual architecture for smart cities.
4. Outline one case study showing how blockchain technology is used in smart cities.
5. Determining and discussing upcoming research issues related to using blockchain technology in smart city applications.

The following sections provide more details on these contributions. This article begins with incorporating blockchain technology in the advancement of smart city areas, which are made up of massive data, the IoE, and the Internet of electricity. It begins by outlining the significance of the IoE, large-scale data, and electricity network frameworks in smart cities while incorporating fundamental ideas, important innovations, and the current state of the blockchain domestically. Next, it lists some of the technical obstacles associated with these three domains and contrasts them with the blockchain. By analyzing the benefits and features of the blockchain, it is possible to see how its technical features are naturally suited for the energy internet and how they can address issues with connectivity and large-scale centers, including the significant expenses associated with maintaining central databases, user privacy breaches, the challenge of upgrading final machinery, and others. A working method is offered to aid with the possible future use of blockchains in smart city applications.

11.1.8 ORIENTATION OF PAPER

The paper unfolds with an introduction that illuminates the theme of sustainable urbanization and transaction processing efficiency, establishing the groundwork for subsequent sections. The conclusion synthesizes key insights, emphasizing the significance of our sustainable urbanization within the existing knowledge landscape. The system model, discussed in Section 11.3, dissects the training, execution, and feedback processes, serving as the conceptual backbone. Section 11.4 translates theory into practice, offering insights into real-world applications, challenges, and results. Section 11.5 navigates open challenges, presenting a proposed sustainable urbanization approach alongside discussions on challenges in transaction processing efficiency and the role of cutting-edge computational resources. The conclusion synthesizes key insights, emphasizing the significance of our sustainable urbanization approach as discussed in Section 11.6.

11.2 RELATED WORK

The exploration of the intersection between IoE data and blockchain within the sharing economy services context has been somewhat limited, despite substantial research efforts in blockchain, sharing economy, and IoE separately. As we look toward the future evolution of collaborative economy services, a significant obstacle arises—the integration of digital progress on the horizon. This convergence is expected to result in the generation of massive amounts of data daily, stemming from the mass crowd and an array of IoE devices.

The bedrock of smart city infrastructure lies in the seamless integration of IoE and cloud computing technologies, forming the backbone of what we know as a smart city. Anchored by the assurance of energy supply through the energy internet, the subsequent utilization of distributed data analysis becomes paramount in guiding city construction and development strategies. Distributed data, in this context, serves as the technical foundation for smart cities, enabling the formulation of scientific city plans and strategic development blueprints [16]. The energy internet, in turn, plays a crucial role in fostering environmentally sustainable urban growth and ensuring the secure and trustworthy consumption of power. In addition, IoE systems introduce unparalleled ease to urban living and administration, with IoE sensors emerging as primary data sources extensively used in smart grids.

11.2.1 Significance of the Internet of Everything

IoE is rooted in its ability to interconnect and interact with objects through information-sensing devices. This not only optimizes resource utilization but also saves time, enhancing the precision and scope of gathering data. Within the framework of smart cities, IoE assumes a key role, in linking intercommunication and connectivity equipment. Equipment with sensing capabilities, in particular, serves as indispensable components, monitoring real-time operations in diverse city scenarios. Looking ahead, it is predicted that sectors, people, workplaces, household devices, machinery, and various objects will be capable of sensing, sharing information, and managing data at any time and from any location [16]. Nevertheless, the task is to develop a framework for smart urban spaces that effortlessly incorporates diverse technologies while maximizing their potential. Moreover, resolutions for smart urban environments should prioritize power efficiency, emphasizing conservation perspectives from both the participants and the environment.

11.2.2 Significance of Distributed Data

Exploring the domain of distributed data, an urban area is a blend of various information categories, classified into organized, partially organized, and unorganized formats. This information encompasses a broad spectrum, including maps data, GPS information, cell phone data, location-based services (LBS) location information, visual surveillance, traffic information, climate and atmospheric data, and ecological information. as well as social interaction information. Starting from a different angle, distributed data analysis greatly helps in various aspects of city life, like making traffic smarter, aiding in decisions of politics, improving healthcare, fostering smart communities, and enhancing tourism [17]. The accumulation of massive dynamic data in cities poses the challenge of extracting meaningful insights and utilizing this valuable information to make cities smarter.

11.2.3 Blockchain and the Fundamental Elements and Challenges in Collaborative Economy through Smart Computing

Transitioning to the realm of smart computing, this field is based on the emulation of human thinking mechanisms, sharing information with various computerized frameworks. Automated advanced learning methods, data analysis, recognizing patterns, and processing natural language are essential components of intelligent computing. Trained smart systems, once equipped with the necessary knowledge, exhibit the ability to function independently without human assistance. In the context of IoE, smart engines serve as powerful brains driving distributed IoE devices [18]. These engines possess the capability to scan through extensive data resources, building intelligence essential for making decisions and future initiatives. Notably, these smart engines play a crucial role in the realm of blockchain, enhancing reliability and privacy concerns.

Within the blockchain framework, smart engines contribute to the reliability and privacy concerns of data from IoE devices by recording and verifying changes in a blockchain ledger. The presence of private key hashing between smart engines and IoE devices further enhances the validation and identification of variations, upholding the data security of the devices. This synergy between smart computing and blockchain not only ensures the reliability of data but also provides a foundation for streamlined and secure, trustworthy transactions. Financial institutions, in particular, have been quick to adopt this technology, facilitating real-time transactions while ensuring trustworthiness through distributed and transparent blockchain ledgers [19]. In this context, smart systems utilize user data to validate and reply to requests in real-time, creating an environment where autonomous blockchain-driven collaborative economy platforms can flourish.

In this automated landscape, monetary transactions and operations are efficiently overseen by smart engines, paving the way for unprecedented security and efficiency. Stakeholders in trusted processes, where the arrangement undergoes examination by the smart platform, and commitments are autonomously completed, eliminating the need for human interaction [20]. A comparison analysis of existing work on smart cities, smart contracts, IoE, and blockchain is represented in Table 11.4. These smart

TABLE 11.4

A Comparison Analysis of Existing Work on Smart Cities, Smart Contracts, IoE, and Blockchain

Topic	Main Focus	Contribution
Smart City	– User-centric security in the city – Provide secure city	– Provide layer architecture, confidentiality, and security in different challenges in smart cities. – Secure application developed for different applications.
Smart Contract	– Deploy smart contract	– Analysis of documentation, manipulation, and justification of security susceptibilities in smart contract.
Internet of Everything (IoE)	– Loopholes in the IoE environment	– Conversation on IoE framework, classification of privacy apprehensions, related limitations and blockchain-based explanations for extenuation safety apprehensions in the IoE.
Blockchain	– Technology provided by blockchain – Security of decentralized technology	– Comprehensive examination of forking, cryptoanalysis, layered construction, consensus, confidentiality, and safety perspective. – Conversation on possible blockchain applications, limitations, and chances. – Review of safety jeopardies, security outbreaks, applicable practical explanations, and upcoming instructions in the blockchain.
IoE with Blockchain	– Blockchain for smart city – Smart contract for smart city – Decentralized application for smart city	– A methodical related work of steeplechases and blockades in a smart city and justification of using blockchain.

contracts, composed of security codes, interact seamlessly with other identical contracts, creating a robust and reliable ecosystem for sharing economy services.

11.2.4 Vehicle Sharing Services

The blockchain serves as a repository for comprehensive vehicle user and vehicle profiles, housing a detailed record of past maintenance, accidents, transfers, and various kinds of immutable data. It establishes connections among the various stakeholders associated with a vehicle through a shared blockchain, contributing to seamless integration in vehicle-sharing economy scenarios. Furthermore, the shared blockchain facilitates instant access to any spontaneous queries related to the vehicle, and the automation of insurance claims is expedited through the application of smart contracts. In Figure 11.2, we depict a scenario in vehicle leasing where data from the IoE, on-board CheckUp(OBCU), and additional vehicle-leasing-related information are sent to a nearby mobile edge node. This node employs blockchain and an

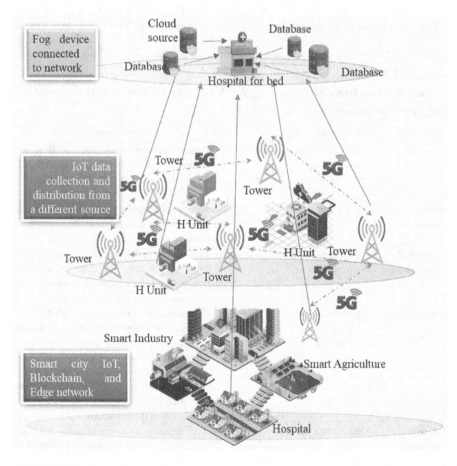

FIGURE 11.2 Scenarios of collaborative consumption for on-the-go car rentals.

off-chain network for storage and later transmits vital information to an artificial intelligence (AI) framework and additional participants in the vehicle ecosystem.

11.2.5 SMART-HEALTHCARE SERVICES

Blockchain technology, combined with the medical IoE, holds the promise of seamlessly connecting diverse stakeholders in spontaneous health-centric sharing economy services [21]. This integration enables the storage of Electronic Healthcare Records (EHR), Electronic Medical Records (EMR), participant profiles, healthcare insurance details, personal healthcare examination records, information about hospital visit frequency, as well as details of healthcare professionals and medical institutions within the distributed and highly protected database. The collaborative system within the healthcare system can achieve firm security and extend beyond conventional boundaries through the integration of blockchain and health-related IoE information.

11.2.6 SERVICES PROVIDED BY THE BLOCKCHAIN SMART CONTRACT

Smart contracts within the blockchain framework can automate the logic of agreements based on location awareness. As an example, a smart contract for health insurance could autonomously integrate policy details, coverage parameters, and affirmation retrieved from health IoE devices.

11.2.7 IDENTITIES MANAGING FRAMEWORK

A significant challenge within the collaborative economy lies in the secure and anonymous management of distinctive identities. Blockchain emerges as a groundbreaking answer by offering securely managing identities, validating users and other publicly available profiles, and providing a digital key signature. This capability establishes a foundation for a universally applicable identity suitable for various collaborative economy frameworks.

11.3 SYSTEM MODEL

The widespread distribution of vast amounts of information across diverse sectors shapes the evolution of forward-looking economic services. The continuous collection, analysis, and utilization of this data enable proactive anticipation, vigilance, and mitigation of potential risks and the active warning of impending situations [21]. This proactive approach helps prevent and mitigate harmful interactions, fostering a more informed and responsive economic environment. The structure enables an IoE node to use distributed messaging platforms, send IoE information of interest with cryptocurrency trading platforms and pathways, save unprocessed IoE-sensed information into a distributed repository through edge networks, and connect with nearby nodes.

The AI system receives information from the IoE, the distributed ledger, and other social media sectors for forensic purposes, sentiment mining, and discovering

TABLE 11.5

Comparative Analysis of Different Applications

Reference	Main Principal	Application	Applied Protocol	Environment
[13]	For the transparent relationship between buyer and seller on product delivery and transaction	Smart contract	NA	e-Commerce
[22]	To develop an efficient and immutable relationship in the market for IoT-based smart city	Origin Chain Ethereum	PBFT	e-Commerce
[19]	The blockchain-based voting system is developed	Fingerprint Authentication	PoC	Smart voting framework
[23]	Propose a system for smart real estate	Smart Contract	PoW	Smart real estate system
[24]	To develop a smart traceable agriculture system	Supply chain	NA	Smart agriculture
[25]	Offer a decentralized system	E-voting system	Longest chain	NA
[26]	To propose a smart connected transportation system	Transportation system	PoM	NA
[27]	Provide a review on sustainable smart city	NA	NA	Smart city

patterns for different shared economic businesses. A comparative analysis of different applications is demonstrated in Table 11.5.

11.3.1 INTERNET DATA SEPARATOR

The components shown in Figure 11.3 gather unprocessed data from IoE gadgets, blockchain, outside-of-the-chain, and diverse internet-based resources. The system pulls resources and features relevant to the collaborative economy and employs a reactive architecture to safely link to third-party application programming interfaces for preparation, caching, and searching. This data acts as the specific marketplace scenario's AI resource.

11.3.2 INTELLIGENT COLLABORATIVE ECONOMY SENTIMENT ANALYZER

This part analyzes, examines multimodal large-scale data, blockchain, and outside-of-the-chain information, gathers semantic numbers, lists the semantic basics, shows the consumer the findings, and modifies the feeling considered to train the framework using either convolutional neural networks (CNN) or deep neural networks models, based on the situation. After the unprocessed information has been cleaned, transformed, and reduced, the neural network algorithm performs semantic information representation by aggregating, classifying, and analyzing predictive information based on the retrieved features [15].

Sentiment-extracting reasoning handles essential aspects of the collaborative economy, such as indexed smart contract phrase information (smart hotels,

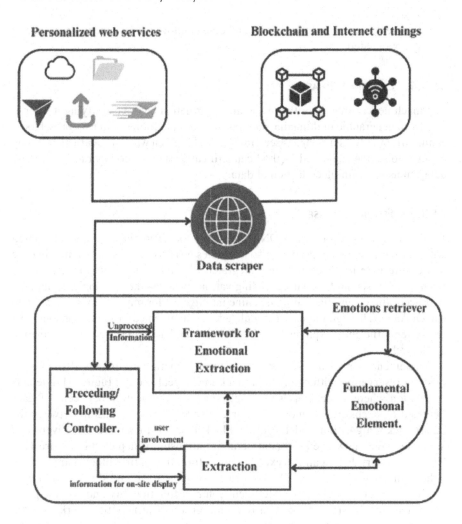

FIGURE 11.3 Blockchain-driven framework for AI support.

autonomous vehicle rentals, immediate doctors, instant hospitals, etc.). Specific APIs have been created for the logic of various primitives, and every one of them uses the structure of the basics as an input to the smart contract logic and as a training dataset. The better the outcomes over time, the richer the dataset becomes. It uses the following three steps to accomplish an intelligent analysis of data.

11.3.3 TRAINING PHASE

In this section, the system attempts to categorize recently arrived satisfied as optimistic, harmful, or impartial by using a supervised knowledge technique. The use of subsequent probability for emotional membership labeling is truly impressive. The way new input samples are mapped, drawing on the features of the sharing economy, is a testament to the innovative approach being employed. This ingenious method is greatly appreciated.

11.3.4 EXECUTION PHASE

The machine learning algorithm evaluates information before classifying it when it gets fresh geographic multimedia material, such as data from social networks, blockchain, IoT, or crowdsourcing, as seen in Figure 11.2. Even when the learned data may contain noise, the suggested method can still categorize the collaborative economy data versus the training collection of data.

11.3.5 FEEDBACK PHASE

While the system incorporates DNN and CNN for leveraging and auto-reasoning with sharing economy datasets, it also integrates NBT to harness human intelligence and ensure user feedback remains integrated. Illustrated in Figure 11.1, an intelligent bot is essential for comprehending various business models' logic, organizing data accordingly, and furnishing machine intelligence for improved visualization to stakeholders. Through increased, simulated, and augmented visualization methods, the system efficiently explores both blockchain on-chain and off-chain information on the web.

Implementation of a collaborative economy operation (IoT-based intelligent hospital reservation with blockchain in a knowledgeable city): Figure 11.4 depicts a situation where an actor uses cyber-physical communication between IoT gadgets and blockchain to request a bed at a hospital. The relationship between the smart lock's algorithm and the other blockchain customers to effectively and safely manage the whole D2D contractual relationship is depicted in Figure 11.5. The IoT smart lock cannot execute and store the entire blockchain ledger. Ethereum virtual machine (EVM) because of its storage and dispensation constraints. Therefore, to achieve intelligent contract functionality and provide proof of settlement, the smart lock must rely on a group of miners that work in a distributed cloud or on edge. As evidence of payment from the entire blockchain, the smart lock will utilize the payment proof; from that, it must recompense the miner blocks that generated the hash. The spatiotemporal terms and conditions of the confirmation are published to the blockchain by the property owner at time g0, as seen in Figure 11.4. At g1, a user looks through the blockchain, accepts the distributed secure contract, and sends money from their digital wallet. At time g2, the smart contract confirms the expense and obligates to the blockchain; however, until the IoE-based smart lock of the stuff of interest is triggered, the expense will be pending.

The prospective tenant contacts the property proprietor at time g4 after a transaction is recorded in the blockchain log. At time g5, the company verifies this contact by requesting the blockchain. At time g7, a tenant receives a connection to the internet smart key card of entry from the landlord after being validated at time g6. After getting it, the patient party can punch the access card to unlock the IoT-based smart secure at time g8. Given its restricted capabilities, the smart lock interacts with the

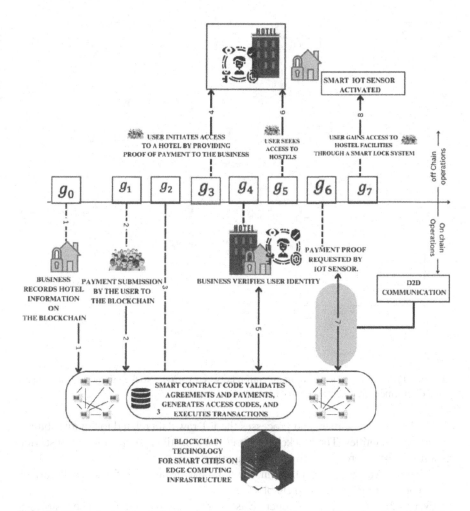

FIGURE 11.4   An instance where a user utilizes blockchain and IoE devices to make a hotel reservation.

MEC node using a distributed incentive-based agreement process, as illustrated in Figure 11.5. Likewise, various distribution economy situations are made to operate automatically and sustain virtual and real-world interactions for large audiences.

11.4   IMPLEMENTATION AND DISCUSSIONS

We set up many permission-confident blockchain nodes at the smartphone edge committed to collaborating on economic operations. The system depends on the consensus in a group of reliable nodes to accomplish scalability and a high volume of collaborative economic operations per hour. Smart contract logic executes the collaborative economy's logic and circumstances, IoE block access strategies, and succeeding information movement.

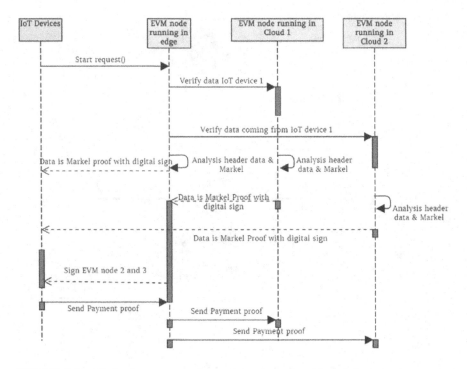

FIGURE 11.5 The intricate communication between various blockchain nodes for transaction commitment in machine-to-machine communication.

The MEC node gathers and processes the IoT raw data related to the distribution economy's amenities. The blockchain stores the essential and prominent transaction data that must be processed to be permanently saved.

Using the Amazon AWS platform, the team has instantiated 50 nodes of various sorts, counting SPV and complete blocks.

Every P2P block is set up to operate as a Linux-simulated engine. The testing set was configured to evaluate the efficacy of smart contracts, various developed APIs, blockchain nodes, remote distributed apps, circulated IPFS-based cloud sources, and terminals leveraging these features. A writing test of four bands was conducted: 1000, 5000, 10,000, and 100,000 transactions. The RAM consumed the time needed to finish the operation. The max operation per block, average operation amount per block, disk consumption, amount of shuts, and whole block generation time were all recorded throughout every transaction.

11.5   OPEN CHALLENGES

This segment delves into numerous pivotal research challenges that hinder the seamless integration of blockchain technology into the frameworks of smart cities. We explore the fundamental cause of these challenges and offer insights for overcoming them, as detailed in the accompanying table [23].

11.5.1 PROPOSED SUSTAINABLE URBANIZATION APPROACH

This strategic framework entails the progressive development of smart cities empowered by blockchain technology, with a steadfast commitment to preserving natural resource deposits. The extensive proliferation of smart IoE devices within urban landscapes has led to a notable surge in the consumption of energy resources. The intricate computational demands imposed by blockchain consensus algorithms, such as PoS, exacerbate this energy demand. Confronting this challenge head-on necessitates the adoption of energy-efficient designs, harnessing renewable energy resources, and implementing energy harvesting as viable solutions. Sustainable blockchain-enabled smart city initiatives demand meticulous considerations, including energy-efficient communication networks, seamless integration of renewable energy resources, efficient storage solutions tailored for blockchain applications, energy-conscious consensus algorithms, and consensus schemes rooted in reputation mechanisms [24]. Notable among these energy-efficient consensus algorithms are alternatives underscored by a central emphasis on minimizing energy usage. Additionally, the incorporation of hardware with relatively lower energy demands, such as energy-efficient processors designed (EEPD) for proof of state-based blockchains, plays a pivotal role in fostering sustainable smart city services. The development of validation schemes that impose less strain on energy resource deposits is imperative for achieving an enduring sustainable approach in a blockchain-based smart city framework, taking into account financial, ecological, and societal perspectives.

11.5.2 CHALLENGES IN TRANSACTION PROCESSING EFFICIENCY

Transaction processing efficiency, encapsulating the time required for transaction execution, and throughput, indicating the volume of transactions within a defined timeframe, play a pivotal role in shaping the scalability of smart cities [25]. In the context of blockchain-enabled smart cities, delays originate from computations at decentralized computing nodes and data communication among nodes. Transaction processing efficiency, encapsulating the time required for transaction execution, and transaction volume within a defined timeframe, plays a pivotal role in shaping the scalability of smart cities. In the realm of blockchain-empowered smart cities, delays arise from computations at decentralized computing nodes and data communication among nodes. Proposing effective solutions to achieve low processing times and enhanced transaction capacity is crucial for the successful integration of blockchain technology in smart cities. Latency resulting from propagation delays in blockchain networks can lead to forking scenarios, where successful miners broadcast their blocks, and others may disseminate theirs before receiving the initial one [25]. This introduces complexities, with miners concurrently possessing multiple blocks, challenging the stability and reliability of the entire blockchain network. As smart cities evolve, addressing these intricacies becomes crucial for optimizing the performance and functionality of blockchain applications within urban environments. In the realm of high-performance computing resource challenges, particularly regarding the surge in smart IoE devices, significant hurdles emerge in terms of elevated storage requirements for providing blockchain-based smart services. To tackle this, utilizing high-performance computing memories with enhanced storage capacity and low power consumption becomes imperative. The integration of these memories at blockchain nodes is viable, while exploring off-network storage solutions introduces security and robustness challenges, demanding meticulous consideration in the strategic design of scalable blockchain-enabled smart cities.

11.5.3 CUTTING-EDGE COMPUTATIONAL RESOURCES

The impending surge in smart IoE devices presents a formidable challenge with heightened storage requirements for delivering blockchain ledger-based intelligent services [26, 27]. Each device within the network must uphold a comprehensive record of transactions, posing limitations on expansiveness. Addressing this challenge necessitates the adoption of high-performance computing memories, characterized by enhanced storage capacity and low power consumption. Beyond integrating these memories into blockchain nodes, exploring off-network storage solutions emerges as a viable avenue. However, the implementation of off-network storage introduces intricate security and robustness challenges, demanding meticulous consideration in the strategic design of scalable blockchain-enabled smart cities [28]. Balancing the imperative need for expanded storage capabilities with the intricacies of ensuring data security and system robustness is a critical undertaking in the evolving landscape of smart cities. As these challenges are navigated, the potential for harnessing the benefits of blockchain technology in the realm of smart services can be maximized, offering sustainable and efficient solutions for the urban environments of the future.

11.6 CONCLUSION

This chapter delves into an analysis of the concepts, attributes, and functionalities concerning the IoE, large-scale data, and the green internet within the framework of a smart city's infrastructure. It examines the notion and distinctive features of blockchain technology while identifying prevailing challenges within the energy internet, IoT, and big data platforms inherent to smart city infrastructure. The study aims to ascertain the compatibility and synergies among their core technologies by juxtaposing their conceptual similarities. Addressing issues such as system security shortcomings, we propose an IoE-based distribution economy scheme that utilizes blockchain frameworks to uphold immutable records. Supported by our projected ML structure, the forthcoming creation of smart cities can furnish cyber-physical distribution economy amenities via IoE data. Through the utilization of smart contracts, this framework can provide intricate spatiotemporal services on a global scale without necessitating central verification authorities. We propose solutions to various challenges, including the high costs associated with constructing and operating large-scale data centers, the inadequate elasticity in defending against attacks, founding trust among green internet users, safeguarding user confidentiality, and adapting the trading market model. Furthermore, we present a resolution to the escalating storage pressure encountered in the latter stages of blockchain development. Our solution advocates for a P2P preserve architecture, intending to offer practical assistance for the future.

REFERENCES

1. Cheng, Jingxian, et al. "Lightweight verifiable blockchain top-k queries." *Future Generation Computer Systems* 156 (2024):105–115.
2. Roy, Deepsubhra Guha, et al. "QoS-aware secure transaction framework for internet of everything using blockchain mechanism." *Journal of Network and Computer Applications* 144 (2019): 59–78.
3. Xiong, Ruoting, et al. "CoPiFL: A collusion-resistant and privacy-preserving federated learning crowdsourcing scheme using blockchain and homomorphic encryption." *Future Generation Computer Systems* 156 (2024): 95–104.
4. Das, Puja, et al. "Block-a-city: An agricultural application framework using blockchain for next-generation smart cities." *IETE Journal of Research* 69.9 (2023): 5773–5783.
5. Ferreira, Célio Márcio Soares, et al. "IoE registration and authentication in smart city applications with blockchain." *Sensors* 21.4 (2021): 1323.
6. Paul, Rourab, et al. "Blockchain based secure smart city architecture using low resource IoEs." *Computer Networks* 196 (2021): 108234.
7. Das, Puja, et al. "Food-Health-Chain: A Food Supply Chain for Internet of Health Things Using Blockchain." *International Conference on Network Security and Blockchain Technology*. Singapore: Springer Nature Singapore, 2023.
8. Jo, Jeong Hoon, et al. "Emerging technologies for sustainable smart city network security: Issues, challenges, and countermeasures." *Journal of Information Processing Systems* 15.4 (2019): 765–784.
9. Almalki, Faris A, et al. "Green IoE for eco-friendly and sustainable smart cities: Future directions and opportunities." *Mobile Networks and Applications* 28.1 (2023): 178–202.
10. Das, Puja, et al. "A Secure Softwarized Blockchain-Based Federated Health Alliance for Next Generation IoE Networks." *2021 IEEE Globecom Workshops (GC Wkshps)*. IEEE, 2021.
11. Ahmed, Imran, et al. "A blockchain-and artificial intelligence-enabled smart IoE framework for sustainable city." *International Journal of Intelligent Systems* 37.9 (2022): 6493–6507.
12. Park, Jin-ho, et al. "CIoE-net: A scalable cognitive IoE based smart City network architecture." *Human-Centric Computing and Information Sciences* 9 (2019): 1–20.
13. Manman, Li, et al. "Distributed artificial intelligence empowered sustainable cognitive radio sensor networks: A smart city on-demand perspective." *Sustainable Cities and Society* 75 (2021): 103265.
14. Das, Puja, et al. "Blockchain-Based COVID-19 Detection Framework Using Federated Deep Learning." *International Conference on Network Security and Blockchain Technology*. Singapore: Springer Nature Singapore, 2021.
15. Tuyls, Robert, et al. "Innovative data-driven smart urban ecosystems: Environmental sustainability, governance networks, and the cognitive internet of everything." *Geopolitics, History, and International Relations* 11.1 (2019): 116–121.
16. Nuraeni, Aisyah, et al. "Smart City Evaluation Model in Bandung, West Java, Indonesia." *2019 IEEE 13th International Conference on Telecommunication Systems, Services, and Applications (TSSA)*. IEEE, 2019.
17. Permana, Asep Yudi, et al. "Smart architecture as a concept of sustainable development in the improvement of the slum settlementarea in Bandung." *International Refereed Journal of Engineering and Science* 2.9 (2013): 26–35.

18. Hayati, Arina, et al. "From smart living into smart city: A lesson from Kampung of Surabaya." *UIA 2017 Seoul world architects Congress.* 2017.
19. Rahman, Md Abdur, et al. "Blockchain and IoE-based cognitive edge framework for sharing economy services in a smart city." *IEEE Access* 7 (2019): 18611–18621.
20. Singh, Saurabh, et al. "Convergence of blockchain and artificial intelligence in IoE network for the sustainable smart city." *Sustainable Cities and Society* 63 (2020): 102364.
21. Swan, Melanie. "Blockchain enlightenment and smart city cryptopolis." *Proceedings of the 1st Workshop on Cryptocurrencies and Blockchains for Distributed Systems.* 2018.
22. Natalia, Naomi Glori, et al. "Bandung City as a smart City model: Navigating policies, challenges, and innovative solutions." *Socius: Jurnal Penelitian Ilmu-Ilmu Sosial* 1.5 (2023).
23. Sharma, Ashutosh, et al. "Sustainable smart cities: Convergence of artificial intelligence and blockchain." *Sustainability* 13.23 (2021): 13076.
24. Anandaraj, A. Peter Soosai, et al. "Convergence of Blockchain and Artificial Intelligence in IoE for the Smart City." *Convergence of Blockchain, AI, and IoE.* CRC Press, 2021. 157–170.
25. Lukić, Ivica, et al. "Possible blockchain solutions according to a smart city digitalization strategy." *Applied Sciences* 12.11 (2022): 5552.
26. Samuel, Omaji, et al. "Towards sustainable smart cities: A secure and scalable trading system for residential homes using blockchain and artificial intelligence." *Sustainable Cities and Society* 76 (2022): 103371.
27. Wong, Phui Fung, et al. "Potential integration of blockchain technology into smart sustainable city (SSC) developments: A systematic review." *Smart and Sustainable Built Environment* 11.3 (2022): 559–574.
28. Khawaja, Sarmad, et al. "Blockchain technology as an enabler for cross-sectoral systems integration for developing smart sustainable cities." *IET Smart Cities* 5.3 (2023): 151–172.

Index

Printed in the United States
by Baker & Taylor Publisher Services